Crystal Growth in Gels

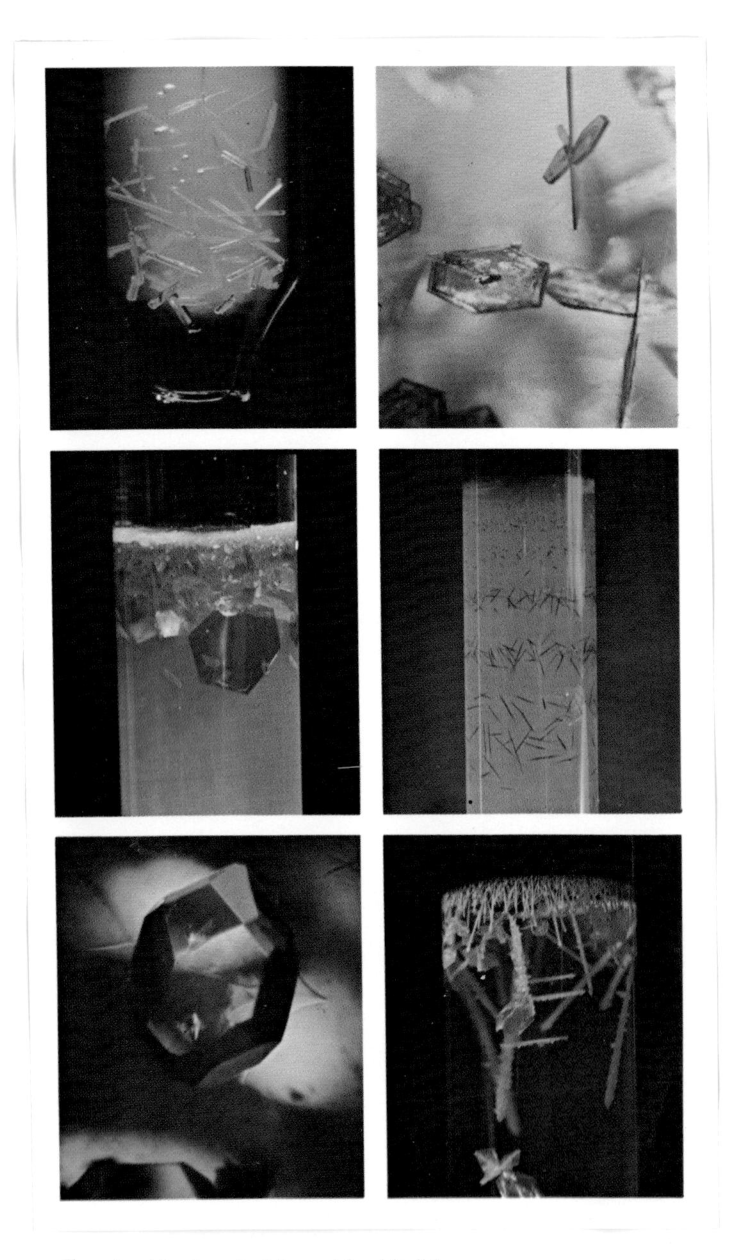

Top: lead hydroxyiodide and lead iodide
Middle: lead iodide and silver chromate
Bottom: calcium tartrate and silver iodide

Crystal Growth in Gels

Heinz K. Henisch

Professor of Applied Physics and Associate Director
Materials Research Laboratory
The Pennsylvania State University

1970

The Pennsylvania State University Press
University Park and London

Library of Congress Catalogue Card Number 77–86379
Standard Book Number 271–00104–6

Printed in the United States of America

Designed by Glenn Ruby

The Pennsylvania State University Press
University Park, Pennsylvania 16802

The Pennsylvania State University Press, Ltd.
London W. 1

To Bridget

Acknowledgments

I am indebted to the late Dr. V. Vand for introducing me to the subject of crystal growth in gels, and to many friends for their collaboration, their illustrative material, their advice and, above all, their unfailing forebearance in the face of enthusiasm: A. F. Armington, G. Bulger, Edna A. Dancy, J. Dennis, J. W. Faust, Jr., E. S. Halberstadt, J. I. Hanoka, K. L. Keester, P. Kratochvil, Sandra McBride, J. W. McCauley, J. Nickl, Janice Perison, R. Roy, C. S. Sahagian, C. Srinivasagopalan and S. K. Suri.

H. K. H.

Contents

Crystal Growth in Gels

1 History and Nature of the Gel Method

1.1 Introduction

It has long been appreciated that advances in solid state science depend critically on the availability of single crystal specimens. As a result, an enormous amount of labor and care has been lavished on the development of growth techniques. In terms of crystal size, purity, and perfection, the achievements of the modern crystal grower are remarkable indeed, and vast sections of industry now depend on his products. So do the research workers whose preoccupation is with new materials, no matter whether these are under investigation for practical reasons or because a knowledge of their properties might throw new light on our understanding of solids in general.

In one way or another, a very large number of new materials have already been grown as single crystals in recent years, some with relative ease and others only after long and painstaking research. Nevertheless, there are still many substances which have defied the whole array of modern techniques and which, accordingly, have never been seen in single crystal form. Others, though grown by conventional methods, have never been obtained in the required size or degree of perfection. All these constitute a challenge and an opportunity, not only for the professional crystal grower but, as it happens, also for the talented amateur. New and unusual methods of growing crystals are therefore of wide interest; and if the crystals are by themselves beautiful, as they so often are, there is no reason why this interest should be confined to professional scientists.

The art and science of growing crystals in gels, largely dormant during the period 1930–1960 and now in the midst of a general renaissance, enjoyed a long period of vogue beginning close to the end of the last century and lasting well into the 1920's. During most of this time, the center of interest was held by the phenomenon of Liesegang Rings. Liesegang was a colloid chemist and a photographer who experimented with chemical reactions in gels (1, 2, 3).* He covered a glass plate with a layer of gelatin impregnated with potassium chromate, and added a small drop of silver nitrate. As a result, silver chromate was precipitated in the form of a series of concentric rings, well developed and with regularly varying spacings. The discontinuous nature of the precipitation, its geometrical features, and the conditions of its occurrence at once became the object of intense, if not altogether successful, investigations. The matter immediately attracted the attention of the great German chemist Ostwald (4) and, in due course, need one mention it, that of Lord Rayleigh (5), thereby receiving what must have appeared as the ultimate seal of respectability. According to a report by Bradford (6), Sir J. J. Thomson, whose principal interests were in very different fields, likewise concerned himself with the problem of periodic precipitation phenomena. (However, the Royal Institution appears to have no record of the 1912 lecture to which Bradford refers.)

Liesegang rings were considered interesting partly because their origin was obscure and partly because they were reminiscent of certain structures found in nature, e.g., the striations of agate. Often they consisted of apparently amorphous material. In due course, the achievement of microcrystalline reaction products also became desirable, because of the ease with which they could be identified by means of x-ray photographs. Larger crystals, several mm in size, were occasionally obtained but not systematically looked for. In contrast, the growth of such crystals is the principal objective of all modern work in this field. An early claim by Fisher and Simons (7) to the effect that "gels form excellent media for the growth of crystals of almost any substance under absolutely controllable conditions" survives as a shining example of faith, no more than a little tarnished by the sporadic nature of its fulfilment to date.

Surprisingly, in view of the history of this subject, the Liesegang ring phenomenon itself is even now only imperfectly understood. It has been displaced from the center of interest and still awaits the talents of a modern Ostwald with an appreciation of its beau-

* References are at the end of each chapter.

ties and with a great deal of time to spare. There is no basic mystery, in the sense that periodic solutions of the diffusion equations (with appropriately chosen boundary conditions) are known to exist in principle, but little is known about the parameters involved and nobody has succeeded in relating theory to practice. Periodic ring and layer formations found in nature offer only the most limited opportunities for research into their origin. Indeed, many are due to very different mechanisms. As one critical analyst has put it when faced with the suggestion that the stripes on tigers and zebras may be glorified Liesegang phenomena: "enthusiasm has been carried beyond the bounds of prudence" (8), a verdict with which the present writer is inclined to concur.

The experiments during the early period derived a good deal of impetus from the interests of geologists, who believed that all quartz on earth was at one time a silica hydrogel. A vein of white gelatinous silica, as yet unhardened by dehydration, was indeed reported to have been found in the course of deep excavations for the Simplon tunnel (9). Moreover, some early experiments were on record, quoted by Eitel (10), according to which microscopic silica crystals had been obtained from silica gels in the presence of various "crystallizing agents" when heated under water vapor pressure. Quite plausibly, then, crystalline foreign deposits found in quartz may be examples of crystal growth in gel. In this way, the gel method appeared to offer systems and opportunities for experiments in "instant geology"(11). Figure 1.1 shows typical examples of natural growth, needles of tourmaline and rutile in single crystal quartz. It is obvious that the needles must have grown first, but current opinion among geologists and mineralogists no longer favors the idea that single crystal quartz is derived from a gel, it now being thought more likely that this particular form is the outcome of hydrothermal growth. Little is known about the viscosity of the hydrothermal medium under the original growth conditions, but it is plausible to suppose that the tourmaline and rutile needles grew in such a medium when it was fluid but viscous. This belief is supported by the knowledge that a gel as such, though generally beneficial, is not absolutely required for the mode of crystal growth under discussion here. On the general subject of the role of gels in geology, there is extensive early literature (12, 13). In the last analysis, the problem of consistency is not yet unequivocally settled as far as quartz is concerned, but there are many examples of crystal growth in other viscous media,

a

b

Fig. 1.1.

Natural growths: needles of (a) tourmaline, and (b) rutile, in quartz.

natural (14) and artificial. Among the unwelcome manifestations of the process are the occasional growth of ice crystals in ice cream, the growth of tartrate crystals in cheese, the crystallization of sulfur in rubber (15), the growth of zinc salts in dry batteries, and, in rapidly descending order of desirability, the growth of uric acid crystals in joints and of stones in human organs (though, of course, these are not ordinarily single crystals). The subject thus has much wider implications than is generally believed, because, among other things, it is distantly related to our understanding of the processes which take place in photographic emulsions.

1.2 Early Work

In the course of early work (e.g., up to 1930) a mass of empirical data was assembled, much of it too imprecise and unsystematic to lead to any real insight into the mechanism of the phenomena involved. However, some of these investigations remain interesting because they have the character of "existence theorems," illustrating at least some of the things that can be done and suggesting new lines of experimentation.

Among the indefatigable enthusiasts was Hatschek (16), working primarily with (5–20%) gelatin and (1–5%) agar gels. He was the first to make a systematic study of particle size distribution in a great variety of Liesegang rings. Among other things, he noted that in ring systems prepared by allowing sodium carbonate to diffuse into a gel charged with barium chloride, distinct and well-formed $BaCO_3$ crystals up to 1.5 mm in length are occasionally found in the otherwise clear space *between* adjacent rings. The observation must have been the first hint, if one was needed, that Liesegang ring formation is indeed a complicated process, especially since Ostwald and Morse and Pierce (17) had earlier placed great emphasis on the sharpness with which the rings are often defined. Moreover, it has since been amply demonstrated that Liesegang rings themselves may consist of substantial crystals (Fig. 1.2a). Even when they do not, the ring systems can be very complex, as shown by the two coexisting periodic structures in Fig. 1.2b. Silver dichromate, and the dichromate, chromate, chloride, iodide, and sulfate of lead, as well as calcium sulfate and barium silicofluoride were included in Hatschek's crystal-growing repertoire. He was also the first to report that crystals grow generally

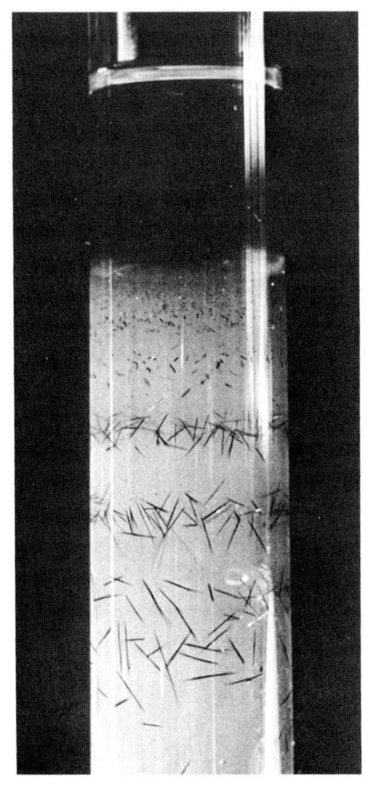

a

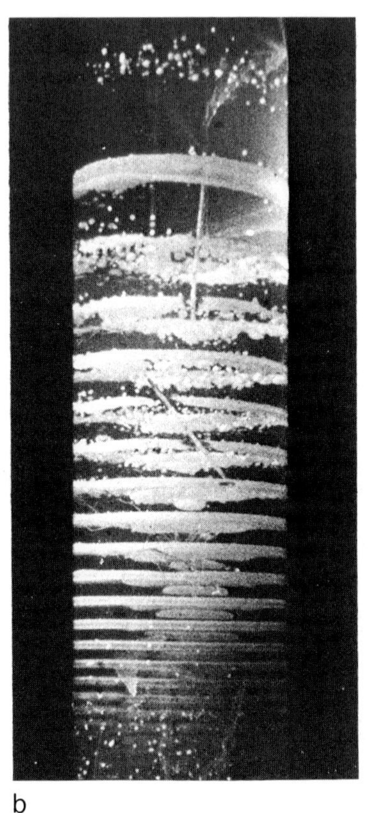

b

better in silicic acid gels than in gelatin or agar. In one of the most truly interdisciplinary experiments on record, Marriage (18) had earlier grown lead iodide crystals in fruit jellies and jams.

It was always realized that reagents could be diffused into a gel, and conversely, that gels could be dialyzed to free them of excess reagents or unwanted reaction products. Such experiments were conducted by Holmes (19), who used the dialyzing process for the treatment of gels in U-tubes in order to eliminate excess reagents which might interfere with the two diffusing components. Holmes also grew single crystals of copper and gold by diffusing a reducing agent into gels charged with the respective salts. Hatschek had done so before, but Holmes claimed "better results than any yet recorded," without actually mentioning crystal sizes. Both experimenters concerned themselves to some extent with the effect of

c

Fig. 1.2.
Liesegang rings: (a) silver dichromate system, (b) calcium phosphate system, (c) gold system.

non-reagent additives such as glucose, urea, and gum tragacanth on crystal growth. Holmes and also Davies (20) noted that reactions in gels can occasionally be influenced by light and, in particular, by short wavelength ultraviolet radiation. Holmes used alkaline gels for the formation of cuprous oxide, but the product was always amorphous.

One of the most important early experiments was performed by Dreaper (21), who wanted to elucidate the role played by the capillarity of the gel structure. For this purpose, he substituted fine sand and even a single capillary tube for the usual gels and found that crystalline growth products could be obtained with such systems. Holmes later used barium sulfate and alundum powders for the same type of demonstration. Strictly speaking, these experiments prove nothing in particular about the nature of gels, but they serve as a valuable hint of the circumstances which favor single crystal formation.

A comprehensive survey of early work on gel structures has been given by Lloyd (22). Structure and classification problems were hotly debated by scientific workers in the 1920's and earlier, often with enviable self-confidence and just occasionally with a trace of venom (23). During the next forty years or so, the subject of crystal growth in gels fell into virtual oblivion, until it was revived by the modern interest in outlandish materials and, more generally, in room temperature growth methods. In comparison, an enormously higher level of interest was sustained throughout this period in Liesegang ring phenomena, a fact for which a bibliography by Stern (24) comprising more than 600 publications bears eloquent testimony. Some beautiful color photographs of Liesegang ring formation have been published by Kurz (25).

1.3 Basic Growth Procedures

To all outward appearances, the gel method is exceedingly simple, but it is now abundantly clear that the physical and chemical processes which determine its outcome are not. One procedure, much used for the simplest demonstration purposes (26), involves the preparation of a hydrogel from commercial waterglass, adjusted to a specific gravity of about 1.06 g/cm^3. This is mixed with an equal volume of approximately 1 molar (1M) acid solution and allowed to gel, a process which depends in a complicated manner on the silicate concentration and on the degree of acidity (see Chapter 2). In the circumstances specified, it generally takes between 24 and 36 hours, after which the gel is left for another 12 hours to allow it to set firmly. Once the gel is formed, some other solution can be placed on top of it without damaging its surface, as shown in Fig. 1.3a. This solution supplies one of the components of the reaction and also prevents the gel from drying out. In order to avoid damage, the supernatant solution should be added dropwise with a pipette, the drops being allowed to fall on the side of the test tube. If the reagent in the gel is tartaric acid and the second reagent is an approximately 1M solution of calcium chloride, then, in due course, crystals of *calcium tartrate* tetrahydrate (only sparingly soluble in water) form in the gel. The first can usually be

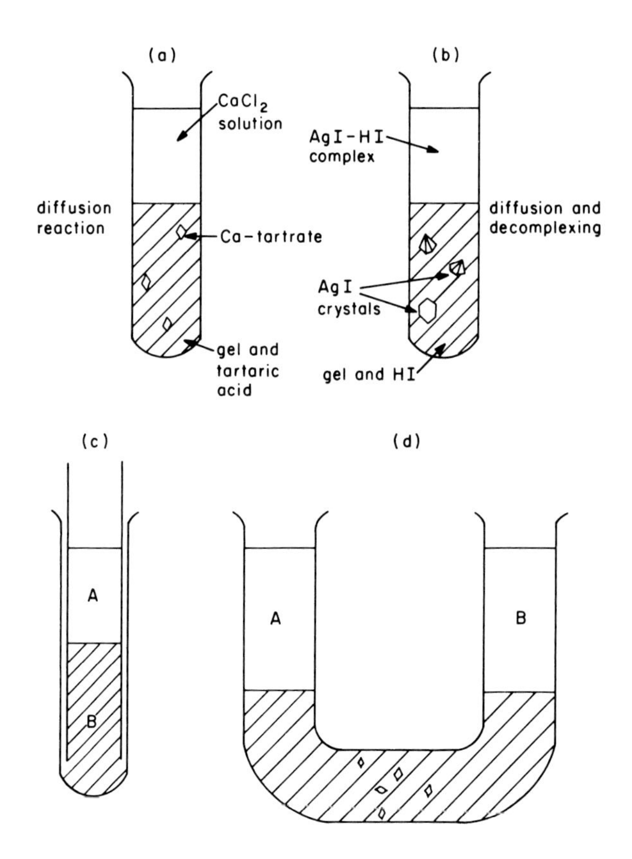

Fig. 1.3.

Basic growth procedures.

a b

seen near the gel surface within an hour or so. Good crystals appear further down the gel column within about a week, and grow to about 8 mm average size. Occasionally, they are much larger but usually less perfect. The speed of formation depends on the concentrations involved, and the time taken before crystals are seen with the naked eye may vary from very few hours to a few days. When much higher acid concentrations are used, sodium tartrate crystals tend to grow in the form of long clear needles.

Because commercial waterglass does not have a fixed and accurately known composition and because it is generally contaminated with undesirable impurities, it is better to use reagent grade sodium metasilicate. A stock solution is prepared by adding 500 ml of water (distilled or demineralized) to 244 g of $Na_2SiO_3 \cdot 9H_2O$. As far as possible, this solution should be kept from contact with the atmosphere to avoid absorption of carbon dioxide (27). As in all cases, the most repeatable results are obtained when the gels are kept under thermostated conditions, e.g., between room tem-

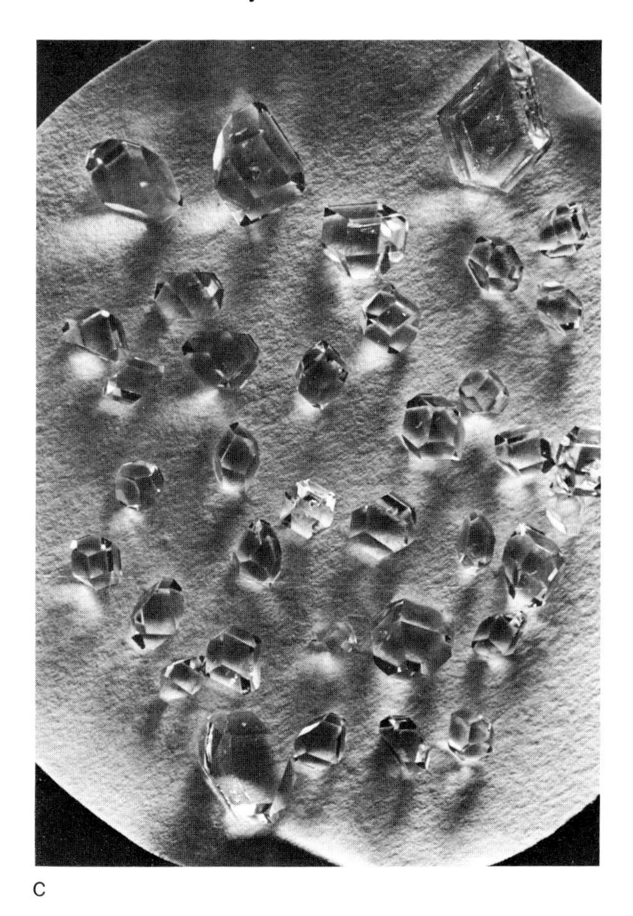

c

Fig. 1.4.

Growth of calcium tartrate crystals: (a) acceptable nucleation; (b) excessive nucleation; (c) gel grown calcium tartrate crystals.

perature and 45 °C, though a really high degree of temperature stability does not seem to be required. Crystals of remarkably high optical perfection can be grown (Fig. 1.4). A more detailed discussion of the factors which govern ultimate size and degree of perfection will be found in Sections 3.2 and 4.5. As a general rule, very dense gels produce poor crystals. On the other hand, gels of insufficient density take a long time to form and are mechanically unstable. A specific gravity of 1.02 g/cm^3 appears to be the lower practical limit.

It is evidently a simple matter to vary the parameters of this system. Thus, calcium tartrate crystals can be grown not only by using $CaCl_2$, but any other soluble calcium salt, such as calcium acetate. However, a straightforward comparison of the process involved is not simple because solutions of equal concentration differ in pH; and, whereas the initial pH can be adjusted, the subsequent changes are not easily monitored and controlled.

A host of other crystals can be grown with varying degrees of success by using different acids and metal salts. Ammonium, copper, cobalt, strontium, iron and zinc tartrates, cadmium and silver oxalates, calcium tungstate, lead iodide, mercuric iodide, calcium sulfate, calcite and aragonite, lead and manganese sulfides, metallic lead, copper, and gold are among those which have been successfully prepared—and there are many others. Nor is it necessary that the second reagent be in the form of a solution. Gas reagents under varying pressures can be used, and they offer the additional possibility of extending the temperature range within which experiments can be carried out. The gel itself need not necessarily be acid (28, 19), nor need it be based on sodium metasilicate; various proprietary silicas ("Ludox," "Cab-O-Sil") can also be used, as can agar gels, though with generally poorer results. The proprietary silicas are free (or relatively free) of sodium, which, in principle, should be an advantage, since it is a contaminant and not part of any essential reaction. On the other hand, gels with and without sodium, but of otherwise identical structures, have not yet been systematically compared. As a practical matter, the point remains unsettled. In a series of early experiments on the growth of lead iodide and lead bromide, Fisher (29) actually found the presence of sodium beneficial for large crystal growth, though it is not clear why.

The formation of calcium tartrate crystals, as described above, is an example of a growth process involving a strong acid as the (unwanted) reaction product, and of course not all processes are of this kind. The growth of *lead iodide,* for instance, does not involve such acid formation. In a typical demonstration experiment, the stock sodium (7.5 ml) metasilicate solution (see above) is diluted with an equal quantity of water. Fifteen ml of 2M acetic acid and 6 ml of 1M lead acetate are then added slowly with continuous agitation. The mixture is allowed to set and 20 ml of 0.75M potassium iodide placed on top of the gel. Good but thin hexagonal plates of PbI_2, about 8 mm in diameter and occasionally larger, grow within about three weeks at 45°C. Lower temperatures (e.g. 20°C) favor greater thicknesses, at the expense of platelet diameter. Crystals up to 3 mm thickness have been grown by Dugan (30). (See also Section 5.1.) When PbI_2 (and other) growth systems are stored for long periods, various changes may take place due to the formation of double salts (29, 31).

Among other lead compounds so far prepared is *lead sulfide* which has been grown in acidic lead acetate gels, covered with a supernatant source of sulfur ions. Brenner and co-workers (32) used dilute solutions (0.1M) of thioacetamide for this purpose. This compound yields sulfur after reaction with the acid which diffuses out of the gel. Cubic crystals of over 1 mm size were obtained. Somewhat smaller crystals (~ 0.5 mm edge) have been reported by Murphy and Bohandy (33), who used Na_2S as the sulfur source.

Calcium tartrate and lead iodide are both grown in initially acid gels, though the PbI_2 system becomes increasingly alkaline in the course of growth. *Lead hydroxy-iodide,* PbI(OH), is an example of a crystal which can (and, indeed, must) be grown in gels which are alkaline from the start. (See also Section 3.4 for growth of calcite and other carbonates.) In this case, the sodium metasilicate and acetic acid solution is adjusted to a pH of 8 by means of KI and allowed to gel. Lead acetate is then diffused into it from a supernatant solution, while the growth system is kept at 40°C or above (28). The crystals which result from this procedure are of the order of 1 mm^2 in cross-section and up to 5 mm long. They are clear, pale yellow in color, and easily distinguished from the more orange hexagonal platelets of PbI_2 with which they coexist in some growth systems (Fig. 1.5).

The processes described above can be influenced electrolytically by the application of electric fields within the gel but, as far as is known, no practical advantage has been derived from this procedure. Mixed crystals can be grown by using mixed reagents, and crystals can generally be doped by the use of small amounts of additives, either in the gel itself or in the top reagent (see below and Section 3.4). By changing the top reagent from one metal salt solution to another, it is sometimes possible to form heterojunctions between isomorphous substances.

In the procedures described thus far, the gel is used as the reaction medium, in which the desired material is chemically formed. In a variant of the method, pioneered by O'Connor and co-work-a

c

b

Fig. 1.5.
Growth of PbI_2 and PbIOH crystals: (a) PbI_2 simple growth, (b) PbI_2 with concentration programming, (c) PbI(OH).

ers (34, 35) for the growth of *cuprous chloride* crystals, the material is first complexed by means of another reagent (HCl) and then allowed to diffuse into a gel free of "active" reagents. Decomplexing sets in with increasing dilution and leads to the high supersaturations necessary for crystal growth. Cuprous chloride is interesting as a possible laser modulator, in which modulation can be achieved by a transverse electro-optic effect (36). The crystals are ordinarily grown from the melt, but thermal strains arise as a consequence of cooling from the melting point at 422°C. In addition, there is a phase change from the wurtzite to the zinc blende structure at 407°C. This is the wanted form, but the cooling process is complicated by this transformation.

In view of these difficulties, the gel method has obvious advantages: the initial experiments described by Armington and co-workers (37) were carried out in 2-cm diameter test tubes containing pH-5 hydrochloric acid gels, with 15 ml of supernatant solution of varying acidity saturated with CuCl. A pH of 5 was found to be optimum. Tetrahedral crystals of 3-mm size grew within about three weeks. As in the case of calcium tartrate, the gel tends to split off the growing crystal, especially at pH values of 6 or higher, resulting in this case in the unwanted growth of numerous microcrystals. In the course of such test tube experiments, conditions in the gel become more acid with time, and crystals formed during an earlier stage of the diffusion process often redissolve later.

Some experiments with U-tubes have also been described, but no practical improvement resulted from their use, except when constant concentration reservoirs were also employed. Best results were obtained (38) with systems which offered not only such reservoirs but also a constant path-length over the cross-section of the gel (Fig. 1.6). This also permitted the gel to be easily removed without breakage for recovery of the crystals or reagent analysis. Of all geometries, the linear column offers the best opportunity for predicting the optimum growth region from a knowledge of the solubility and the variation of pH with distance. Armington and O'Connor (39) have shown how such estimates can be obtained, and as long as the crystals are reasonably small compared with the diffusion cross-section, the agreement with experiment is good. The use of reservoirs and linear column doubled the size of the crystals as compared with those grown in test tubes. Moreover, by diffusing into a shorter gel column, the growth process was con-

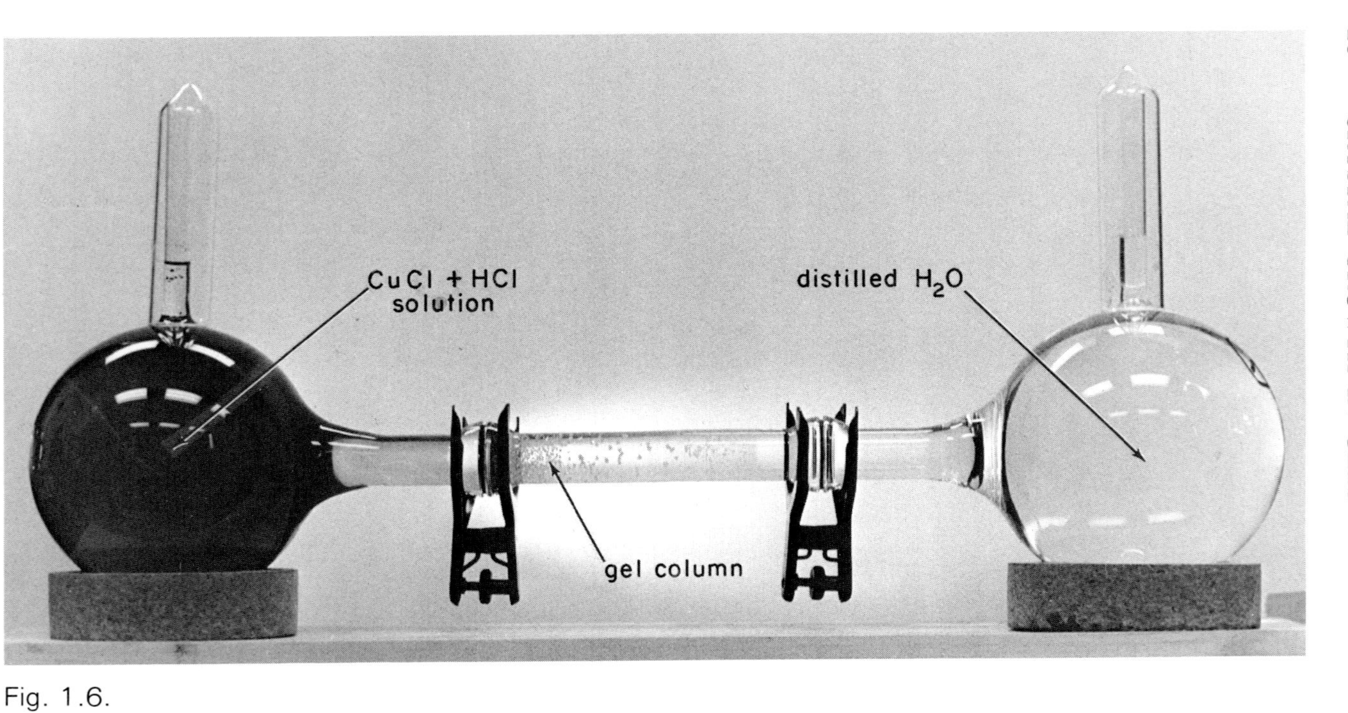

Fig. 1.6.
Double diffusion system with reagent reservoirs and straight diffusion column; column diameter 20 cm, length 25 cm. [After Armington and O'Connor (38, 39)].

siderably speeded up. Special analytic procedures were developed for monitoring the progress of the decomplexing process (40). Of the large number of crystals grown, between 5 and 10% were optically clear. These were entirely free of silicon in the bulk, though a small amount remained in the surface even after cleaning. Surface cleaning is best achieved by means of HF (1 part of 48% solution to 10 parts of H_2O), followed by rinsing in hydrochloric acid and acetone (37). Storage in vacuum is recommended. In air the crystals may become discolored due to the formation of $CuCl_2$ on the surface. Cuprous chloride can alternatively be grown by the interdiffusion of cupric chloride and hydroxylamine hydrochloride, as mentioned by Torgesen and Sober (41), but the method has not yet been optimized to the same extent. Though fewer experiments have been made, it is known that *cuprous bromide* may be grown by analogous procedures (38).

In a manner similar to that described above, *silver iodide,* which has always proved difficult to grow, can be complexed with potassium iodide (Fig. 1.3b). The great solubility of AgI in KI solution and its rapid decrease with increasing dilution makes this method particularly suitable. The first successful experiments on growth in solution were reported by Cochrane (42) and in gels by Halberstadt (43). For the latter, gels were prepared from 7 ml sodium silicate solution (244 g $Na_2SiO_3 \cdot 9H_2O$ + 500 ml water), 8 ml 2M potassium iodide, and 15 ml 2M acetic acid. The gelling mixture was allowed to set in a 25 x 200 mm test tube at 45°C; and a solution made from 57 g silver iodide, 260 g potassium iodide, and 250 ml water was added carefully when the gel had set. This solution was allowed to diffuse into the gel for about a week, poured off and replaced by water. The temperature was kept at 45°C throughout. Pale yellow hexagonal plates appeared within a few hours and grew to about 10 mm diameter. One side of these crystals was smooth, the other showed ridges along the diagonals between corners of the hexagon and along lines parallel to the sides of the hexagon, suggesting that growth occurs along only one direction of the c-axis. In the course of electron microprobe analysis, no contaminating silicon from the gel was found (to an accuracy of a few p.p.m.), but potassium was found in quantities up to 6 per cent in certain regions of the crystals, chiefly along the valleys between the ridges.

Smaller hexagonal plates (about 3 mm) can be obtained by a slightly different and shorter procedure. Gels are prepared as described, except that potassium iodide solution weaker than 2M, or water, is used in making the gelling mixture. The addition of the supernatant AgI-KI complexed solution on top of the gel resulted in the growth of these smaller crystals within a week. Similar experiments at room temperature (about 23°C) with gels initially free of KI give somewhat different results, inasmuch as crystals grow in the aqueous layer above the gel as well as in the gel. In Halberstadt's experiments, the crystals in the gel appeared as completely clear small pyramids and prisms (about 0.05 mm). The crystals in the aqueous layer were hexagonal pyramids, 5 mm in diameter and 5 mm high. They had good surfaces but were translucent rather than transparent.

In another variant of the AgI growth procedure, HI may be used as the complexing agent in place of KI, provided oxygen is excluded. For reasons which remain obscure, the presence of HI favors the formation of clear pyramids, as opposed to hexagons. The complex may be gradually diluted or else poured off after a time and replaced by water. The former procedure is obviously capable of sustaining growth for a longer time. The addition of

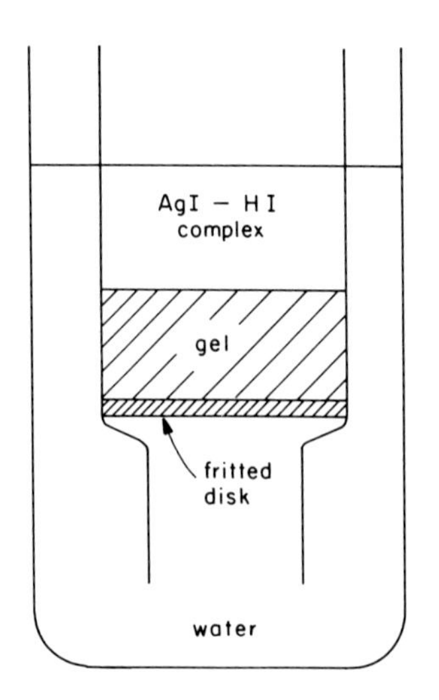

Fig. 1.7.

Double diffusion system with fritted disk [After Nickl and Henisch (51)].

supernatant complex may be preceded by an intermediate stage during which 2M HI is diffused into the gel to lower the local pH and thereby to raise the solubility of AgI. This has a small but significant effect in lowering the number of crystals formed at the immediate gel interface (44).

The system geometries (Fig. 1.6) employed for CuCl can also be used for AgI, as can the modified double diffusion method shown in Fig. 1.7. Figure 1.8 gives examples of AgI growth systems, and Fig. 1.9 shows some of the single crystals grown. The best are absolutely clear and of gem-like quality. Sizes up to 5 mm base diameter have been achieved. In refreshing contrast with calcium tartrate, AgI is of considerable interest, partly in connection with experiments on the mechanism of cloud seeding and partly as an ionic conductor.

Silver bromide and *silver chloride* have been grown by Blank and co-workers (45) in similar ways, AgBr being complexed with HBr and AgCl with NH_4OH. Decomplexing (of a PbI_2 solution in KI) is also an alternative method of growing single crystal lead iodide. The decomplexing procedure is thus a versatile and valuable addition to the repertoire of gel growth methods, and the geometry employed permits the removal of waste products in a simple and continuous way.

It is also possible to grow crystals by dissolving microcrystalline material in, say, an acid and to allow the solution to diffuse into a gel medium of a pH at which the solubility is much lower. *Antimony sulfur-iodide,* an interesting ferroelectric material, is an example. It may be formed, for instance, in concentrated hydriodic acid containing Sb_2S_3 when the pH is raised by dilution or partial neutralization (46). (See also Section 4.4.) Another variant involves the diffusion of alcohol into a gel containing the material to be grown. *Triglycene sulfate,* another ferroelectric crystal, serves as an example for this procedure (47). The material is highly soluble in water but much less so in alcohol. A gel is made of sodium metasilicate stock solution (see above), and mixed with equal parts of glycene (14M) and sulfuric acid (2M). Supernatant methanol diffuses into this gel after setting and produces bulky crystals and needles of triglycene sulfate, often with excellent surfaces and optical clarity. The needles may be about 1 mm thick and up to 1 cm long (Fig. 1.10). A similar growth process in solution (without gel) has been used by Nitsche (48). (See Appendix for growth of metallic copper and lead.)

a

Fig. 1.8.

Silver iodide growth systems: (a) and (b) undoped, (c) copper doped [After Henisch and Suri].

As noted above, the simplest test tube configuration is not always the best. On occasions, it is useful to employ a double tube, as shown in Fig. 1.3c. The gel may be withdrawn by means of the inner tube, from which it can then be ejected in the form of a continuous column by gentle air pressure. In other circumstances, a double diffusion system is desirable, either in the shape of a U-tube (Fig. 1.3d), or as a concentric configuration. Such arrangements are actually necessary for reagents which would form an immediate precipitate when mixed with un-gelled sodium metasilicate solution. Because double diffusion systems involve more variables, they are somewhat harder to optimize than the simpler

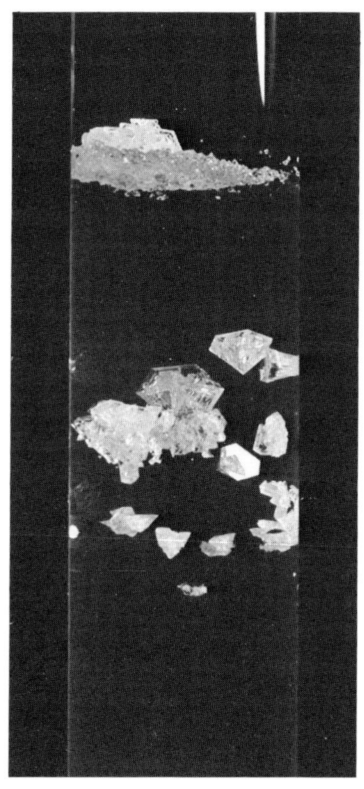

b

c

arrangements. In principle, not only double but multiple diffusion systems can be devised, e.g., for doping purposes or for the production of mixed crystals.

There is evidence that the crystals grown in gels can be considerably purer than the reagent used would lead one to expect, a fact which may be connected with the low growth temperature. Armington and co-workers (37) have documented the purification effect for Fe, Pb, Cr, Al, and Ca in CuCl and for Mg, Ni, Na, and Sr in calcium tartrate (49). Conversely, enrichment effects are also observed, e.g., as described by Dennis and Henisch (50) for iron in calcium tartrate, and by Nickl and Henisch (51) for vari-

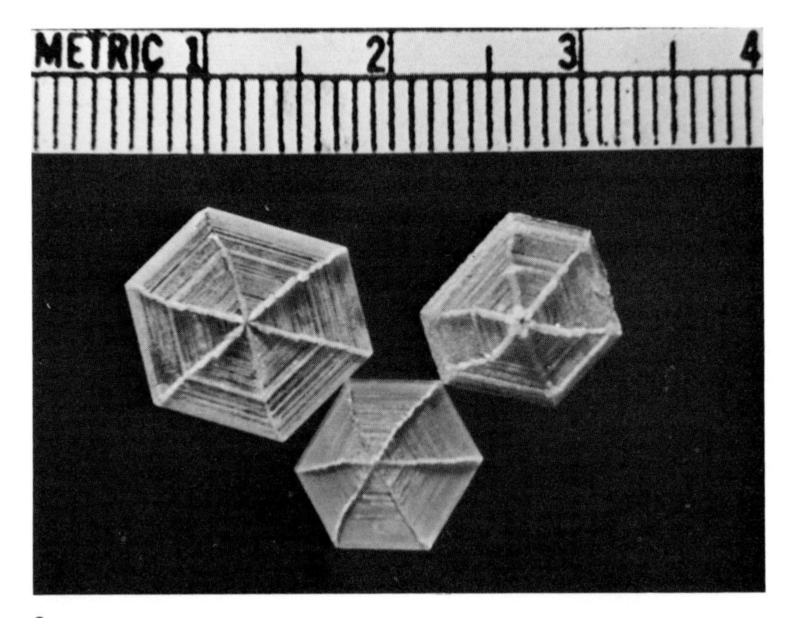

a

b

Fig. 1.9.

Gel-grown silver iodide crystals: (a) hexagons grown by dissociation of the AgI-KI complex, (b) pyramids grown by dissociation of the AgI-HI complex. 4 mm base diameter [After Halberstadt and Suri (44)].

Fig. 1.10.

Growth of triglycene sulfate [After Perison (47)].

ous dopants in calcite. All these matters are very much in need of further investigation; for the moment there is no safe criterion which enables one to predict whether a given dopant (or contaminant) will be eliminated from or enriched within the growing crystal. The absence, under all normal conditions, of any appreciable silicon contamination may be plausibly ascribed to the general stability of the gel network. However, there are cases (notably calcite) which do not conform to this rule, and these must be dealt with by the hybrid methods described in Section 3.4.

Doping can be easily achieved by adding the dopant to the gel or supernatant reagent in solution form. In this way, a variety of doped crystals may be produced. Ni, Cr, Fe, Co and Nd can, for instance, be incorporated into calcium tartrate. In all these cases, the crystals become colored, yellowish green, deep green, yellow, pink, and purple, respectively. To some extent, the color is a function of doping level. Thus, 0.065 atomic percent of neodymium have been reported to make the crystals faintly blue, whereas 1.3 atomic percent make them purple. Contrary to expectations, the doped crystals were not fluorescent, probably due to absorption of the host lattice. When present in small amounts, the dopants do not affect the outward appearance of crystal perfection (see also Section 3.1). However, without special measures, completely uniform doping density within any crystal is not ordinarily achieved. When the dopant diffuses into the gel together with the second reagent, its average concentration in the crystal is, of course, a function of this distance from the gel interface. The incorporation of neodymium into calcium tartrate has been studied in some detail (49). If Nd^{+3} replaces Ca^{+2}, charge compensation is required if appreciable amounts of dopant are to be accommodated. Sodium ions can serve this purpose. One would expect one Na^{+1} ion to be incorporated (in place of Ca^{+2}) for every neodymium ion. Whether this is actually what happens remains to be checked. At low doping levels (only), Armington and co-workers have reported uniform neodymium distributions.

Many of the systems described above lend themselves readily for demonstrations, but large and near-perfect crystals are not generally achieved without at least some degree of optimization of the experimental parameters. This is a highly rewarding, generally lengthy and, at least for the present, largely empirical process, though various useful guidelines will be found in the chapters which follow. Moreover, it is clear that the crystals in any given system grow competitively, and it is therefore necessary to limit nucleation, an aspect which receives special discussion in Chapter 4. In the absence of nucleation control, there is a greater chance statistically of finding a really large crystal in a large growth system than in a small one. Inasmuch as the walls of the container constitute a growth impediment, they reinforce this tendency. Armington and co-workers (49) were able to grow calcium tartrate crystals of ⅜″ maximum size in 50 ml test tubes, and crystals of 1″ length and about ½″ diameter in 500 ml containers.

In all cases and irrespective of procedure, the nature of the reaction product should be determined by direct analysis, since double salts and hydrated compounds are often formed. For some mysterious reason, unexpected crystals seem to grow more easily than expected ones, and unwanted crystals best of all! Estimated doping levels should be approached with the same degree of caution.

For a better understanding of the processes described above, it is necessary to have some insight into the nature and structure of gel media. These matters are discussed in Chapter 2.

References: Chapter One

1. Liesegang, R. E. *Phot. Archiv.* 221 (1896).
2. ________. *Chemische Reaktionen in Gallerten.* Düsseldorf (1889).
3. ________. *Colloid Chemistry* (Ed. J. Alexander). 783. Chemical Catalog Co., New York (1926).
4. Ostwald, W. *Z. Phys. Chem.* 27:365 (1897).
5. Lord Rayleigh *Phil. Mag.* 38:738 (1919).
6. Bradford, S. C. *Colloid Chemistry* (Ed. J. Alexander). 790. Chemical Catalog Co., New York (1926).
7. Fisher, L. W., and Simons, F. L. *Amer. Mineralogist.* 11:124 (1926).
8. Hedges, E. S. *Colloids.* Edward Arnold Co., London (1931).
9. Spezie, G. *Atti. Acad. Torino.* 34:705 (1899).
10. Eitel, W. *The Physical Chemistry of Silicates.* Univ. of Chicago Press (1954).
11. Holmes, H. N. *J. Phys. Chem.* 21:709 (1917).
12. Cornu, A. *Kolloid Zeitschrift.* 4, 5 (1909).
13. Krusch, C. *Z. prakt. Geol.* 15:129 (1907); *Z. prakt. Geol.* 18:165 (1910).
14. Koide, H., and Nakamura, T. *Proc. Imp. Acad. Japan.* 19:202 (1943).
15. Endres, H. A. *Colloid Chemistry* (Ed. J. Alexander). 808. Chemical Catalog Co., New York (1926).
16. Hatschek, E. *Kolloid Zeitschrift.* 8:13 (1911).
17. Morse, H. W., and Pierce, G. W. *Z. phys. Chem.* 45:589 (1903).
18. Marriage, E. *Wied. Ann.* 44:507 (1891).
19. Holmes, H. N. *Colloid Chemistry* (Ed. J. Alexander). 796. Chemical Catalog Co., New York (1926).
20. Davies, E. *J. Amer. Chem. Soc.* 45:2261 (1923).
21. Dreaper, G. *J. Soc. Chem. Ind.* 32:678 (1913).
22. Lloyd, D. J. *Colloid Chemistry* (Ed. J. Alexander). 767. Chemical Catalog Co., New York (1926).

23. von Weimarn, P. P. *Colloid Chemistry* (Ed. J. Alexander). 27. Chemical Catalog Co., New York (1926).
24. Stern, K. H. *Bibliography of Liesegang Rings.* National Bureau of Standards Miscellaneous Publication No. 292 (1967).
25. Kurz, P. *Science and Technology.* March issue, 81 (1965).
26. Henisch, H. K.; Dennis J.; and Hanoka J. I. *J. Phys. Chem. Solids.* 26:493 (1965).
27. Halberstadt, E. S. University of Reading, U.K., personal communication (1967).
28. Dennis, J.; Henisch, H. K.; and Cherin, P. *J. Electrochem. Soc.* 112: 1240 (1965).
29. Fisher, L. W. *Amer. J. Science.* 15:39 (1928).
30. Dugan, A. E. Pennsylvania State University, personal communication (1967).
31. Fisher, L. W., and Simons, F. L. *Amer. Mineralogist.* 11:200 (1926).
32. Brenner, W.; Blank, Z.; and Okamoto, Y. *Nature.* 212:392 (1966).
33. Murphy, J. C., and Bohandy, J. *Bull. Amer. Phys. Soc.* 12:327 (1967).
34. O'Connor, J. J.; DiPietro, M. A.; Armington, A. F.; and Rubin, B. *Nature.* 212:68 (1966).
35. Armington, A. F., and O'Connor, J. J. *Mat. Res. Bull.* 2:907 (1967).
36. Murray, L. A. *Electronic Industries.* 23:83 (1964).
37. Armington, A. F.; DiPietro, M. A.; and O'Connor, J. J. Air Force Cambridge Research Laboratories (Reference 67–0445), Physical Sciences Research Paper No. 334 (July, 1967).
38. ———— and O'Connor, J. J. *Proc. Int. Conf. on Crystal Growth.* Birmingham, U. K., July 15–19, 1968; *J. Crystal Growth* 3, 4:367 (1968).
39. ———— ————. *Mat. Res. Bull.* 3:923 (1968).
40. O'Connor, J. J.; Thomasien, A.; and Armington, A. F. Air Force Cambridge Research Laboratories (Reference 68–0089), Physical Sciences Research Paper No. 352 (February, 1968).
41. Torgesen, J. L., and Sober, A. J. National Bureau of Standards Technical Note No. 260, p. 13, May (1965).
42. Cochrane, G. *Brit. J. Appl. Phys.* 18:687 (1967).
43. Halberstadt, E. S. *Nature.* 216:574 (1967).
44. Suri, S. K. Pennsylvania State University, personal communication (1969).
45. Blank, Z.; Speyer, D. M.; Brenner, W.; and Okamoto, Y. *Nature* 216: 1103 (1967).
46. Dancy, Edna A. Westinghouse Research Laboratories, personal communication (1969).
47. Perison, J. Pennsylvania State University, personal communication (1968).

48. Nitsche, R. *Helv. Phys. Acta.* 31:306 (1958).
49. Armington, A. F.; DiPietro, M. A.; and O'Connor, J. J. Air Force Cambridge Research Laboratories (Reference 67–0304), Physical Sciences Research Paper No. 325 (May, 1967).
50. Dennis, J., and Henisch, H. K. *J. Electrochem. Soc.* 114:263 (1967).
51. Nickl, J., and Henisch, H. K. *J. Electrochem. Soc.* 116:1258 (1969).

2 Gel Structure and Properties

2.1 Gel Preparation and Properties

Although it is true that good crystals can occasionally be grown in substances which are not normally classified as gels, the general observation is that gels and, in particular, silica gels, are the best and most versatile growth media. Their preparation, structure, and properties therefore deserve attention. At the same time, it is useful to note that no clearcut demarcation lines between gels, sols, colloidal suspensions, and pastes have ever been established. Standard descriptions of these materials are certainly available but they are not nearly as crisp as one would wish and many practical substances must be regarded as borderline cases. A gel, for instance, has been defined as "a two-component system of a semi-solid nature, rich in liquid" (1), and no one is likely to entertain illusions about the rigor of such a definition.

The materials which are ordinarily called gels include not only silica gel (e.g., as usually prepared from sodium metasilicate solution), but also agar (a carbohydrate polymer derived from seaweed), gelatin (a substance closely related to proteins), soft soaps (potassium salts of higher fatty acids), a variety of oleates and stearates, polyvinyl alcohol, and various hydroxides in water. Closest to gels in structure are sols, which are likewise two-component systems, but resemble liquids more than solids. There are also hybrid media which consist of small jelly-like particles separated by relatively large tracts of liquid phase. These are sometimes called "coagels." In other cases (e.g., hydrated strontium sulfate), the gels appear to consist of crystalline needles in bundles (2). Resistance to shear can be used as a semi-quantitative criterion for the comparison and classification of gel materials. Because crystals can grow in a variety of gels and gel-like media, precise differentiations may not be very important in the present context.

Gels are formed from suspensions or solutions by the establishment of a three-dimensional system of cross-linkages between the molecules of one component. The second component permeates this system as a continuous phase. A gel can thus be regarded as a loosely interlinked polymer. When the dispersion medium is water, the material should be called a "hydrogel," to distinguish it from the brittle solids which are often obtained by subsequent drying (e.g., "silica gel"). In practice, the distinction is not always made; the meaning is usually clear from the context.

The gelling process can be brought about in a number of ways, sometimes by the cooling of a sol, by chemical reaction, or by the addition of precipitating agents or incompatible solvents. Gelatin is a good example of a substance which is readily soluble in hot water and can be gelled by cooling, provided that the concentration exceeds about 10%. In smaller concentrations, the mixture remains a quasi-liquid, the number of cross-linkages being evidently insufficient to establish a recognizable gel. In a similar way, non-aqueous gels can be prepared by cooling sols of aluminum stearate, oleate, or naphthenate in hydrocarbons. On the other hand, Alexander and Johnson (1) quote some substances (e.g., certain cellulose nitrates in alcohol and methyl celluloses in water) which actually show the opposite behavior: they gel on being warmed. Gelling by the addition of incompatible solvents or precipitating agents is likewise a simple matter. Solutions of ethyl cellulose, cellulose aceto-stearate, or polystyrene in benzene, for instance, can be gelled by rapid mixing with ether, in which the substances are less soluble. Similarly, a gel of dibenzol-cystine can be prepared by dissolving the material in alcohol and pouring the solution into water (3). Gels of aluminum and ferric hydroxides, vanadium pentoxide, and bentonite can be made from aqueous suspensions by the addition of suitable salts, e.g., $MgCl_2$, $MgSO_4$, or KCl. Although gelatin gels (see above) do not need such additions, their firmness and transparency depend on the pH of the solution before gelling and on the nature of the ions present. A variety of practical gel preparation "recipes," e.g., for oil, nitrocellulose, aniline nitrate, and pectin gels, have been given by Bartell (4). Most gels are mechanically and optically anisotropic, except when under strain. However, according to Thiele and Micke (5), the presence of high ion concentrations can bring about the formation of unstrained nonisotropic gels by the alignment of nonspherical sol particles.

Gels can also be formed by the action of two reagents in concentrated solution, e.g., barium sulfate from barium thiocyanate and manganese sulfate. This is the type of process which gives rise to silica gels and which, for present purposes, is the most important. Detailed procedures for the preparation of a whole series of silica-alumina gels of varying pore size have been described by Plank and Drake (6, 7). In these cases, the aluminum salt is dissolved in some acid before mixing with waterglass or sodium metasilicate. Transparent gels are formed within a few minutes. They have complicated ionic adsorption properties and are ordinarily used as cracking catalysts. When crystals are grown in them, they tend to be contaminated with aluminum (8), which is why they are not ordinarily used for this purpose, despite their otherwise attractive flexibility.

The gelling process itself takes an amount of time which can vary widely from minutes to many days, depending on the nature of the material, its temperature, and history. For silica gel this has been described and documented by Treadwell and Wieland (9). During a prolonged period after mixing ("incubation time"), the liquid hydrosol remains outwardly unchanged. This is followed by a comparatively rapid and pronounced increase in viscosity and, in due course, by quasi-solidification. Even before this stage is reached, standard viscosity measurements become meaningless because the material is non-Newtonian. Since gelling is a matter of degree, quoted gelling times are always very approximate. The mechanical properties of fully developed gels can vary widely, depending on the density and on the precise conditions during gelling. For instance, silica gels with a molecular silica-to-water ratio of 1:30 or 1:40 can easily be cut with a knife. At 1:20 the medium is rather stiff, and at 1:10, friable (10). Still denser silica gels have conoidal fracture surfaces similar to glass. The internal surface area likewise depends on the detailed circumstances during gel preparation (11). The available information on silica gels is somewhat "informal," but the mechanical properties of gelatin gels, relevant as they are to applications in photography, have been investigated a good deal more systematically and have a substantial literature of their own (2, 12–15).

It is often reported that reagents diffuse as rapidly through gels as through water, an observation which dates back to 1862, but it has long since been found that this is true only for electrolytes and very dilute gels. It is certainly not true for large molecules (e.g., organic dyes; see below), nor for colloids. One operative parameter is obviously the size of the diffusing particles, relative to the pore size in the gel. Another is the amount of interaction (if any) between solute and internal gel surfaces. Because convection currents are never completely absent in liquid (gel-free) systems, precise comparisons are very difficult. Stonham and Kragh (16) studied the diffusion of KBr through gelatin gels and reported a linear diminution of diffusion coefficient with gel concentration, but no systematic dependence of diffusion on mechanical gel properties. Kurihara and co-workers (17) concluded from their experiments on the diffusion of the sulfate ion in gelatin gels that the effective diffusion constant is, in this case at any rate, controlled by surface absorption. Corresponding experiments on silica gels do not appear to have been carried out. There is no evidence that the diffusion constant of small atoms and ions is sensitively influenced by the silica gel density, as long as that density is low. This makes it plausible to conclude that it is not greatly influenced by the presence of the gel at all.

In the course of experiments on crystal growth, the need to determine reagent concentrations in gels arises frequently, and the adsorptive properties of gels make this a difficult problem. A variety of substances adsorb on silica hydrogel with particular ease. It has been shown (18), for instance, that 1 molecule of iron oxide can be adsorbed on every 5 molecules of silica, and that thorium and yttrium form particularly strong bonds (19). The chemisorption of alkali ions has been studied by Köppen (20) and cannot fail to be of some importance, considering that such gels are ordinarily made from sodium metasilicate (see Section 2.2). There is also a selective adsorption of organic dyes; stains produced by fuchsin, methyl violet, and malachite green, for instance, cannot be removed by dialysis, whereas acidic dyes can be leached out. In some cases, there is a change of dye color on adsorption, and this has been ascribed to a polarizing effect of the gel surface. There is room for many theories, but even greater is the need for more precise and systematic experimentation. What is known to date is sufficient to amount to an emphatic caution: the *total* and *free* solute contents of a gel may not be the same. We have as yet a very imperfect understanding of the role played by the internal gel surface, but it is unlikely that its function can be ignored.

2.2 Gelling Mechanism and Structure of Silica Hydrogels

When sodium metasilicate goes into solution, it may be considered that monosilicic acid is produced, in accordance with the dynamic equilibrium

$$Na_2SiO_3 + 3H_2O = H_4SiO_4 + 2NaOH,$$

and it is generally accepted that monosilicic acid can polymerize with the liberation of water:

$$\mathrm{HO{-}\overset{\overset{OH}{|}}{\underset{\underset{OH}{|}}{Si}}{-}OH} + \mathrm{HO{-}\overset{\overset{OH}{|}}{\underset{\underset{OH}{|}}{Si}}{-}OH} \rightarrow \mathrm{HO{-}\overset{\overset{OH}{|}}{\underset{\underset{OH}{|}}{Si}}{-}O{-}\overset{\overset{OH}{|}}{\underset{\underset{OH}{|}}{Si}}{-}OH} + H_2O.$$

This can happen again and again until a three-dimensional network of Si-O links is established, as in silica:

$$\begin{array}{l} \quad\;\; \mathrm{OH} \quad \mathrm{OH} \\ \quad\quad\; | \qquad\;\; | \\ \mathrm{HO{-}Si{-}O{-}Si{-}}\ldots \\ \quad\quad\; | \qquad\;\; | \\ \quad\quad\; \mathrm{O} \qquad \mathrm{O} \\ \quad\quad\; | \qquad\;\; | \\ \mathrm{HO{-}Si{-}O{-}Si{-}}\ldots \\ \quad\quad\; | \qquad\;\; | \\ \quad\;\; \mathrm{OH} \quad \mathrm{OH} \end{array}$$

As the polymerization process continues, water accumulates on top of the gel surface, a phenomenon known as *syneresis*. Much of the water is believed to have its origin in the above condensation process, and some may arise from purely mechanical factors connected with a small amount of gel shrinkage.

Figure 2.1a shows that the time required for gelation is very sensitive to pH. Because gelation is a gradual process, there is no unique definition of gelation time, but almost any definition will serve for comparative tests when linked with a standard procedure. Hurd and Letteron (21) have described such a method based on mechanical gel properties, and Alexander (22) described one based on measurement of the reaction rate with molybdic acid. The results on Fig. 2.1a agree with those given by other workers (23) and suggest convincingly that the reaction is ionic in character (contrary to the impression conveyed by the simplified representation used above). Some of the electrochemical subtleties (but alas not all) have been instructively and charmingly described in a semipopular book by Alexander (24). There is much that is still unclear about the mechanism, but it is known that two types of ions are in fact produced: $H_3SiO_4^-$ and $H_2SiO_4^=$, in relative amounts which depend on the hydrogen ion concentration. The latter, which is favored by high pH values, is in principle more reactive, but the higher charge implies a greater degree of mutual repulsion. $H_3SiO_4^-$ is favored by moderately low pH values and is held to be responsible (7) for the initial formation of long-chain polymerization products. In due course, cross-linkages are formed

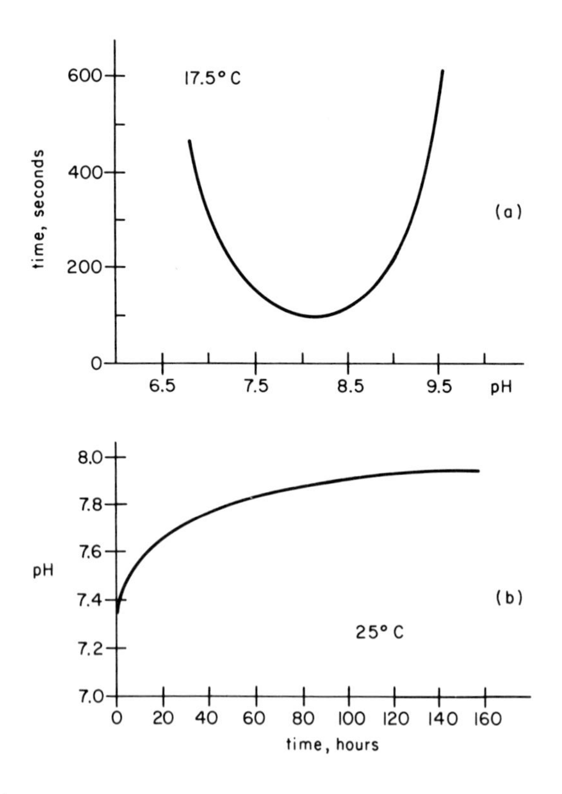

Fig. 2.1.

Gelling process and acidity; sodium metasilicate, in the absence of other reagents: (a) effect of pH on gelling time, (b) pH changes of an originally neutral gel during syneresis [After Planck (7)].

between these chains, and these contribute to the sharp increase of viscosity that signals the onset of gelation. It is plausible to assume that very long chains, because of their lower mobility, will cross-link more slowly than short chains. At very low pH values, the tendency towards polymerization is diminished, and chain formation is slowed. There is thus a complicated interplay of reactions which, at any rate in the absence of other reagents, leads to a minimum gelling time at a pH of 8. A really precise interpretation of this minimum is not yet available. It appears to be strongly temperature dependent; Alexander (25) reports that at 1.9°C gelling is almost instantaneous at pH 6. There is also some evidence that different mechanisms predominate below and above pH 3.2. Figure 2.1b shows how the pH of an initially neutral gel increases during syneresis, probably as a result of the progressive and stabilizing hydroxyl substitution for oxygen in the polymerized structure (7):

$$-\overset{|}{\underset{|}{Si}}-O^- + H_2O \rightarrow -\overset{|}{\underset{|}{Si}}-OH + OH^- .$$

In the presence of other ions, the pH changes can be very different (26). All this means that an initial pH measurement before gelling can be misleading as a guide to the prevailing acidity of any gel growth system in the near-neutral range. In the high and low pH ranges, the effect is not likely to be very important. The formation of cross-linkages can be encouraged by the partial substitution of Al for Si. Because of the difference in valency cross-links form more easily. Gelling time is reduced and the resulting gels (when subsequently dried) have a higher density and a smaller pore size than those without aluminum (6).

The well-known stability of the silicon-oxygen bonds is responsible for the fact that the above polymerization process is largely irreversible. By way of an exception, Hurd and Thompson (27) have reported on dilute silica hydrogels made with waterglass and acetic acid which could be liquefied by shaking, but it is not clear whether the resulting liquid was Newtonian or not, and there is the possibility that only some of the bonds were actually ruptured. The structure of the silicic acid-water system has been extensively investigated in many other ways, and several comprehensive reviews are available which, in turn, give references to earlier work (10, 28, 29). By way of comparison, gelatin gels are believed to be bonded by much weaker forces, possibly of the van der Waals type. Such gels can generally be re-liquefied by heating even without mechanical agitation. In the course of gelling, most systems become somewhat opaque; indeed, light-scattering experiments have been used as an alternative to the methods given above for studying the kinetics of the gelling process (30, 26).

The above representation must not be taken to mean that the structural network in a silica hydrogel has uniform pores. In fact, it shows only their minimum size, since closed rings of larger size, with and without cross-linkages, can easily be formed. It is therefore reasonable to suppose that there is a distribution of pore sizes within each gel, and that one gel is distinguished from another by the nature of this distribution. In the course of some early experiments on gels containing radium or radiothorium, Biltz (31) concluded that hydrogels are characterized by two types of pores, "primary" pores of nearly molecular dimensions and much coarser "secondary" pores which behave as more or less normal capillaries. However, whether such a sharp distinction can be made must remain open to doubt. In the ordinary way, even the largest pores are too small to be seen under the microscope, though structures are occasionally shown uo by ultramicroscopy. Under x-rays, the patterns of silica hydrogel are the same as those observed for silica glass, except for the more intense background which indicates that the gel is less homogeneous (3, 32). By observing the progressive sharpening of the interference lines, structural changes which take place during gel aging can be followed (e.g., see 33, 34, and 35).

It is an interesting fact that when a gel contains bubbles, these are usually lenticular in shape, even though they may begin as spheres in the sol. Their orientation is often parallel to one another which suggests at least some degree of long-range order, unless it can be shown to be due to external factors, e.g., the direction of the maximum temperature gradient. The matter does not seem to have been investigated for silica gels, but Hatschek (36) has studied it in gelatin by diffusing acetic acid into gels containing sodium carbonate. He found that the bubbles become lenticular quite suddenly during the setting process, an interesting point which deserves to be followed up.

It is generally difficult, though not impossible, to separate the liquid phase by mechanical means, but water extraction by means of drying agents (e.g., concentrated sulfuric acid) has been known for a very long time, e.g., van Bemmelen (37). The outcome depends, among other things, on the water vapor pressure (drying rate) and on the previous gel history (38). Drying has only a small effect on volume, a fact which demonstrates the relative rigidity of the network structure. Dried silica gels can be more readily subjected to x-ray analysis and other investigational procedures appropriate for solids. As a result, a good deal more is known about them than about the hydrogels from which they are made; and because the fractional volume change on gentle drying is small, measurements on dried gels may yield results which are significant for hydrogels also. On the other hand, the changes which occur during extensive drying (e.g., heating to 1000°C) are irreversible and must thus involve the structural network as well as the liquid phase. With this caution in mind, it is possible to perform a gravimetric determination of nitrogen absorption (6, 7, 39) and to calculate the internal surface area from the absorption isotherms. This, in turn, leads to the establishment of models of the internal structure and to estimates of the average pore size. Such estimates are necessarily dependent on assumed pore geometries. They yield *effective* pore diameters of the order of 50–160Å for silica gels and of 28–35Å for silica-alumina gels of varying density. In this way, it has also been possible to show that syneresis (if allowed to precede water-dialysis and drying) has little effect on the average pore size.

There is no doubt that the basic gel structure affects the crystal growth characteristics, including growth rates and ultimate crystal size. The detailed mechanisms involved in growth will be described in Chapter 3, the effect of gel pore size on nucleation in Section 4.5.

References: Chapter 2

1. Alexander, A. E., and Johnson, P. *Colloid Science*. Vol. 2, Clarendon Press, Oxford (1949).
2. Hedges, E. S. *Colloids*. Edward Arnold & Co., London (1931).
3. Lloyd, D. J. *Colloid Chemistry* (Ed. J. Alexander), 767, Chemical Catalog Co., New York (1926).
4. Bartell, F. E. *Laboratory Manual of Colloid and Surface Chemistry*. Edwards Brothers, Ann Arbor, Michigan (1936).
5. Thiele, H., and Micke, H. *Kolloid Z.* 111:73 (1948).
6. Plank, C. J., and Drake, L. C. *J. Colloid Science*. 2:399 (1947).
7. ———. *J. Colloid Science*. 2:413 (1947).

8. Perison, J. Pennsylvania State University, personal communication (1968).
9. Treadwell, W. D., and Wieland, W. *Helvetica Chim. Acta.* 13:856 (1930).
10. Eitel, W. *The Physical Chemistry of Silicates.* University of Chicago Press (1954).
11. Madelay, J. D., and Sing, K. S. W. *J. Appl. Chem.* 12:494 (1962).
12. Saunders, P. R., and Ward, A. G. *Nature.* 176:26 (1955).
13. ________. *Rheology of Elastomers.* Pergamon Press, London (1958).
14. Johnson, P., and Metcalf, C. J. *J. Photogr. Science.* 11:214 (1963).
15. Ward, A. G. *Brit. J. Appl. Phys.* 5:85 (1954).
16. Stonham, J. P., and Kragh, A. M. *J. Photog. Science.* 14:97 (1966).
17. Kurihara, H.; Higuchi, H.; Hirakawa, T.; and Matuvra, R. *Bull Chem. Soc. Japan.* 35:1740 (1962).
18. Müller, K. K. Dissertation, Technische Hochschule, Stuttgart (1939).
19. Sahama, T. G., and Kanula, V. *Ann. Acad. Scient. Fennicae 42A.* No. 3 (1940).
20. Köppen, R. *Kolloid Zeitschrift.* 89:219 (1938).
21. Hurd, C. B., and Letteron, H. A. *J. Phys. Chem.* 36:604 (1932).
22. Alexander, G. B. *J. Amer. Chem. Soc.* 75:5655 (1953).
23. Hurd, C. B.; Raymond, C. L.; and Miller, P. S. *J. Phys. Chem.* 38:663 (1934).
24. Alexander, G. B. *Silica and Me. Anchor Books* (Doubleday), Garden City (1967).
25. ________. *J. Amer. Chem. Soc.* 76:2094 (1954).
26. Greenberg, S. A., and Sinclair, D. *J. Phys. Chem.* 59:435 (1955).
27. Hurd, C. B., and Thompson, L. W. *J. Phys. Chem.* 45:1263 (1941).
28. Hauser, E. A. *Silicic Science.* Van Nostrand, Princeton, N. J. (1955).
29. Iler, R. K. *The Colloid Chemistry of Silica and Silicates.* Cornell University Press, Ithaca, N. Y. (1955).
30. Audsley, A., and Aveston, J. *J. Amer. Chem. Soc.* 84:2320 (1962).
31. Blitz, M. *Z. phys. Chem.* 126:356 (1927).
32. Warren, B. E. *Chem. Reviews.* 26:237 (1940).
33. Krejci, L., and Ott, E. *J. Phys. Chem.* 35:2061 (1931).
34. Holtzapfel, L. *Kolloid Z.* 100:386 (1942).
35. Böhm, J. *Kolloid Z.* 42:276 (1927).
36. Hatschek, E. *Kolloid Z.* 49:244 (1929).
37. van Bemelen, K. *Z. anorg. Chem.* 20:265 (1902). There are many other relevant papers in the same journal, going back to 1878.
38. Anderson, J. S. *Z. phys. Chem.* 88:191 (1914).
39. Deryagin, B.; Friedlyand, R. F.; and Krylova, V. *Doklady Akad. Nauk. SSSR.* 61:653 (1948).

3 Growth Mechanisms and Characteristics

3.1 Diffusion and Growth Rates; Functions of the Gel

Gels are obviously not impermeable, but the fact that convection currents are suppressed, above a certain magnitude at any rate, can easily be demonstrated. With an ordinary microscope it is possible to verify that particles have streaming and Brownian motions in the ungelled solution but are at rest after gelling. With a laser-ultramicroscope arrangement of the kind described by Vand, Vedam, and Stein (1), this demonstration can be extended to smaller particles, e.g., down to about 600Å and even below, depending on the wavelength and intensity of the laser light. Such tests do not rule out the possibility of convection currents on a submicroscopic scale, but it is implausible that these play any major role.

In the absence of convection, the only mechanism available for the supply of solute to the growing crystal is diffusion. One may envisage that the solute supersaturation ϕ_∞ at large distances from the crystal remains unchanged during growth. At the crystal surface ϕ would initially have the same value but would then adjust itself in the course of growth to the lower value ϕ_0. This would be determined by the dynamics of the growth process itself. It is interesting to check the extent to which this notion can be verified by measurements on growing crystals as a function of time, even though the verification is bound to be indirect.

Frank (2) has developed equations which give a description of diffusion-controlled growth rates for several different idealized geometries. They are intended to apply to systems in which the growth rate is limited only by volume diffusion and not in any sense by processes which occur at the crystal surface itself. In ordinary growth systems, stagnancy is hard to achieve but in view of the observations described above, gels are a near-ideal medium for experiments under such conditions. Since the essential correctness of the Frank model in terms of concentration contours has already been confirmed by means of multiple-beam interferometry (3), it is reasonable to make use of the model in the present context (4). The growth rates calculated by Frank involve the "reduced radius" (5), which, for a spherical system, is defined as $r/(DT)^{-1/2}$, where r is the radius of the crystal, D the diffusion constant, and t, time. The theory yields a simple result in the form

$$\phi_\infty - \phi_0 = F(s). \tag{3.1}$$

By measuring s and knowing the function F, the value of ϕ_0 at any time could be determined in principle. What is more important in the present case, a constant value of s implies a constant value of ϕ_0, as long as D does not change. The constancy of s can, of course, be checked by plotting r against $t^{-1/2}$ or, since the zero point of the time scale is now known, by plotting r^2 against t (5, 6).

Some limitations must be borne in mind: one arising from the initial transient period during which steady-state concentrations are established, and one arising from exhaustion of the available solute. Both factors must be expected to give rise to nonlinearity. The duration of the first is not easy to calculate, but the second can be estimated without difficulty from the original reagent concentrations in the gel and from the distance between the crystal under observation and its nearest neighbor. In addition, one might expect difficulties due to departures from the simple geometry envisaged by Frank. However, different linear projections used for r have in practice given consistent results. There remains, in the general case, substantial uncertainty as to the effect which the disruption of the gel structure (especially during the initial stages of crystal growth) has on the local value of D. This applies also to the effect of pH changes which occur during growth.

Through the techniques described in Section 1.3, several types of crystals have been grown, some in systems which involve only simple crystallization in the presence of the gel and some which involve a chemical reaction to produce the solute. Linear r^2 vs t relationships have been observed in both cases, suggesting that the chemical reaction was not the rate-determining process in crystal

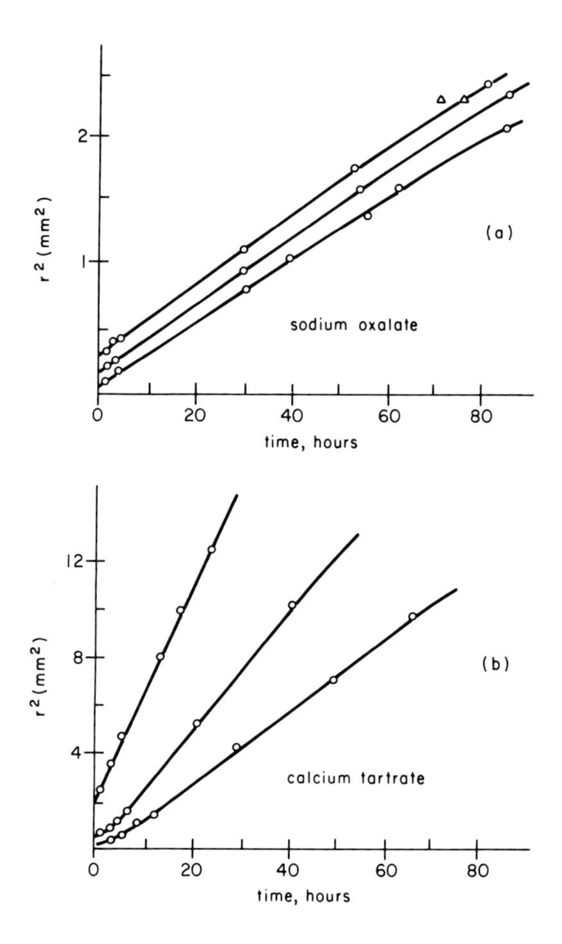

Fig. 3.1.

Crystal size as a function of time; typical relationships [After Henisch, Hanoka and Dennis (4)].

growth. Results are shown in Fig. 3.1. In each case, departures from linearity set in during the later stages of growth, at times which were plausibly associated with the onset of interaction between the diffusion regions surrounding neighboring crystals. The appearance of distinct linear regions may be taken as a confirmation of the constant surface-supersaturation hypothesis. Indeed, it is remarkable that the parabolic relationship holds true, not only

Fig. 3.2.

Growth veils near center of gel-grown calcium tartrate crystals.

for bulky crystals which permit an approach to a spherical diffusion pattern, but also for needles. Faust (7), for instance, has found the same equation to hold for the length of thin dendrites of metallic lead (for growth times between 10 and 200 minutes). Rate constants may be evaluated from the slopes and their behavior studied under various growth conditions.

The fact that the supersaturation ϕ_0 is self-adjusting to the needs of the growth process may have something to do with the high degree of crystalline perfection observed; it is one of the distinctive features of the gel method. It would, of course, be very interesting if the detailed parameters of these systems could be determined by local microanalysis of the medium in the immediate vicinity of the growing crystal. In principle, another possibility would be the optical determination of concentration contours, but light scattering by the gel makes this difficult.

Most calcium tartrate crystals exhibit growth veils near their geometrical center. Examples are shown in Fig. 3.2. These veils are evidently formed during the initial stages of growth and may well be associated with the nonlinearities near the origin on Fig. 3.1b. Such features are not peculiar to crystal growth in gels; they are frequently observed in the course of growth from solution. From independent evidence, Egli and Johnson (8) have ascribed them to "a growth rate temporarily greater than the crystal can

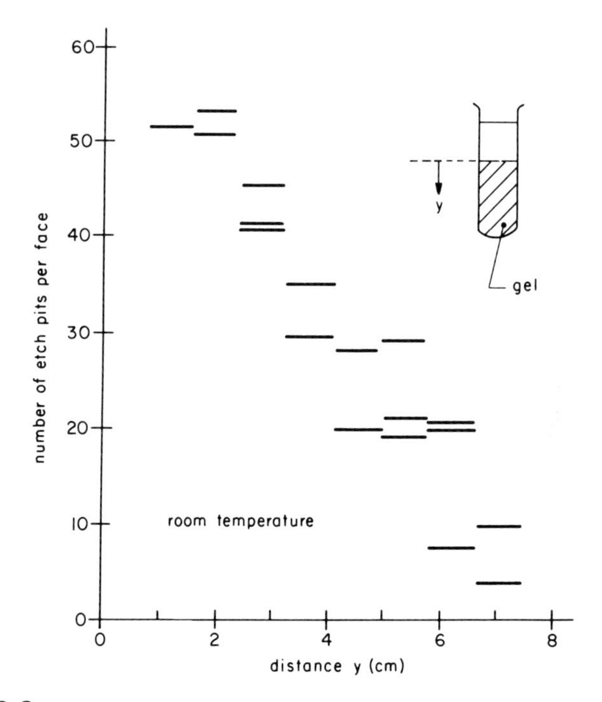

Fig. 3.3.

Etch pit count as a function of distance from the diffusion interface; calcium tartrate crystals of approximately equal size [After Henisch, Hanoka and Dennis (4)].

tolerate"; and whereas this is less explanatory than one might wish, it is in harmony with the comments above. It is also in agreement with results obtained by Armington and co-workers (9) in the course of doping experiments: crystals growing rapidly take up more impurity than those growing slowly.

In systems which depend on the diffusion of one reagent through a gel charged with another, the average crystal growth rate is greatest near the top of the diffusion column where the concentration gradients are high, and smallest near the bottom, where the gradients are also small. Corresponding to this distribution of growth rates, there is also a distribution in the number of etch pits on any crystal face, as shown for calcium tartrate in Fig. 3.3. If the occurrence of etch pits is accepted as a measure of disorder, in the most general terms, then Fig. 3.2 and the arguments above suggest strongly that the growth rate itself determines the number of defects built into the crystal, even in the absence of foreign impurities. Such a process had already been envisaged by Jagodzinski (10) and Washburn (11) and presumably is linked with the low surface mobility of molecules at the growth temperature. The average etch pit densities observed on gel-grown calcium tartrate crystals are of the order of 10^3–10^4 pits/cm^2. Occasionally, specimens are found which have etch pit densities smaller by an order of magnitude. The high degree of perfection has also been demonstrated by means of Laue transmission patterns (9).

Once new solute has been brought to the surface by diffusion, growth takes place either via screw dislocations (after the initial stages of growth; at any rate, see Section 4.2) or via two-dimensional surface nucleation. Figure 3.4 shows the surface of a calcium tartrate crystal with characteristic growth layers which spread successively across the face from one central point. The process of getting solute molecules to the active growth points is probably governed by surface diffusion. The surface diffusion coefficients are expected to increase (11) with increasing temperature, leading to greater perfection and thus to fewer etch pits. A small but significant effect of this kind has been observed on calcium tartrate over the practical range of growth temperatures (20°–65°C), though there are other, and possibly more potent, factors which would tend to give the same results (see Section 4.6). It is quite generally true that the crystals which grow in the lower regions of growth systems are larger than those which grow near the gel interface, but proof of greater perfection (given above for the case of calcium tartrate) has not yet been provided for other crystals. Such information would be valuable, even though, in the last analysis, the features arising from smaller diffusion gradients and those arising from less competitive growth (see below) are not easily distinguished.

It has been demonstrated by de Haas (12) that crystals can in principle be grown by the straightforward diffusion of a gaseous reagent into a solution. Without gel, however, the resulting specimens were smaller than those ordinarily obtained by gel methods, presumably because nucleation could not be suppressed, nor could convection be altogether avoided.

The establishment of a stable pattern of concentration gradients as discussed above is regarded as one of the principal functions of the gel. The gel acts, moreover, as a "three-dimensional crucible"

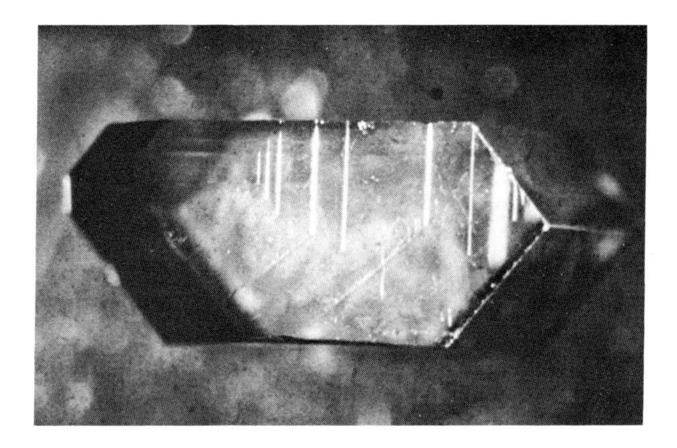

Fig. 3.4.
Surface of a growing calcium tartrate crystal.

which supports the crystal and, at the same time, yields to its growth without exerting major forces upon it. This (relative) freedom from constraint is believed to be an important factor in the achievement of high structural perfection. Indirect supporting evidence for this view comes from doping experiments. The softness of the gel and, presumably, the uniform nature of the forces which it exerts upon the growing crystal make it possible to overdope specimens until they are metastable as a result of severe internal strain. This can be demonstrated (6), for instance, by adding a nickel salt to a calcium tartrate growth system. Ni is accommodated in the crystals, which become green as a result but remain perfectly clear. When the amount of Ni exceeds a certain level, the crystal "explodes" on contact, audibly if not violently. Ni contents of the order of 1% certainly produce this effect which may conceivably have research applications. Of course the growing crystal must do some work against the surrounding gel, and a few simple attempts have been made to measure "crystallization pressures" (13).

Important as the above functions may be, the principal role played by the gel is that of suppressing nucleation, thereby reducing the competitive nature of the growth. Indeed, nucleation control is believed to be the key to the ultimate success of the gel method. The problems and opportunities involved are discussed in Chapter 4. The growth mechanisms operative after nucleation should not be very different from those in stagnant solutions, though complications may arise in the cases in which the gel serves both as a reaction medium and as a diffusion medium.

3.2 Ultimate Crystal Size; Re-implantation

In all simple gel-growth systems (i.e., those without constant concentration reservoirs), the crystals reach a stable, ultimate size and it is easy to see why this must be so. In fact, two mechanisms are simultaneously at work, and the ultimate crystal size may be determined by either. On the one hand, there is the progressive exhaustion of the reagents, which may be simply demonstrated by cutting a cylindrical growth system into disk-like layers and by performing the necessary analysis on each layer. Figure 3.5 shows the results of such an analysis for the case of calcium tartrate. In the gel region close to the supernatant liquid, the tartaric acid content is completely exhausted, partly through the formation of calcium tartrate crystals and partly through a demonstrable loss of tartaric acid to the solution on top. As the diffusion process continues,

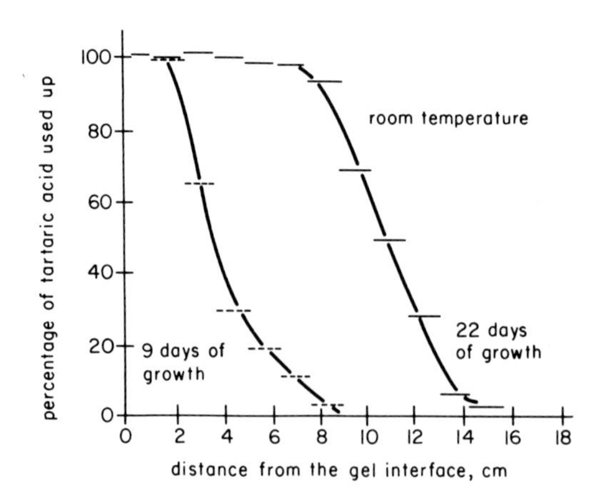

Fig. 3.5.

Calcium tartrate growth; analytic profile of gel column [After Dennis (6)].

the reaction boundary moves down the column, as the results show. At the same time, the diffusion gradients generally diminish (though in Fig. 3.5 not very obviously), a tendency which is enhanced, of course, by the fact that the tubes are of finite length.

Hand-in-hand with the diffusion of principal reagents, there is a diffusion of hydrogen and hydroxyl ions which governs the local pH. The form of the pH distribution may be very different from the concentration profiles shown in Fig. 3.5. Reagent exhaustion can (and often does) reduce the speed of growth to levels which amount to zero-growth for practical purposes.

The second stabilizing factor arises from pH and related considerations. In cases involving the salts of weak acids, of which calcium tartrate is an example, there is a major change towards lower pH values in the course of crystal growth, since the reaction yields rapidly diffusing hydrogen ions. In the increasingly acid environment (down to pH 1), calcium tartrate is increasingly soluble and a steady state may eventually be reached in which, without any shortage of reagents, growth and solution are in balance. Similar considerations apply to the growth of calcite from (say) $CaCl_2$ and Na_2CO_3. In cases like PbI_2, there is no such acid formation, but PbI_2 is increasingly soluble in the growth medium as the concentration of (alkaline) potassium acetate increases. This is one of the reaction products, and the fact that it is alkaline favors the formation of $PbI(OH)$ as a secondary consequence (see Section 1.3). This is why thin pale yellow needles of this compound are sometimes found near the bottom of old PbI_2 growth systems.

The demand for large crystals must come to terms with the question: how large is large *enough,* and the answer depends, of course, on the nature of the experiments contemplated. By means of modern investigational techniques, a great deal can be learned about very small crystals—for most laboratory purposes, crystals of a few millimeters in length are large enough. The problem of growing much larger crystals resolves itself into the problems of limiting nucleation (Chapter 4), of ensuring a continued reagent supply, and of removing the waste products. Reservoirs and continuous flow systems can solve the supply problem. Waste product removal is most easily achieved for the decomplexing procedures described in Section 1.3, in which one end of the diffusion column is in permanent contact with distilled water. Corresponding provisions could be made for other growth procedures, though not in the same straightforward way.

When such procedures for promoting growth are unavailable or inconvenient, it is still possible to re-implant crystals from an exhausted gel into a new one. In the course of such a transfer, crystals must be handled with great care to minimize surface damage. One method is to place such a crystal onto the surface of a set gel in a tube and to cover it with more sodium metasilicate solution, which is then allowed to set. Before adding the supernatant reagent, the temperature of the system may be raised temporarily to permit a (possibly damaged) surface layer of the crystal to dissolve. The boundary between new gel and old gel tends to support more nucleation than the gel volume as such. To reduce this problem, the old gel surface should be protected from dust while exposed to the air. With these precautions, crystals can be re-implanted repeatedly (5) and can be increased in size during each stage. For calcium and copper tartrates this has been done (14) up to four times, with maximum (total) weight increases by factors of 19. A statistical spread of these factors is always observed. Similar experiments have been made with calcite, but the method has not otherwise been widely applied, and it remains to be seen how versatile it is.

The criterion for a successful re-implantation is, of course, the degree of order at the boundary between old and new growth. This is very much influenced by the procedure used. Etching before growth promotes orderly boundaries. It can be performed in two ways: outside or inside the gel (before the addition of new supernatant reagents); or with or without simultaneous heating (see above). For reasons discussed in Section 3.3, etching *in situ* is the more successful process. In calcium tartrate, at any rate, it can lead to growth boundaries which defy visual detection. Imperfect boundaries, when produced, are contaminated by inclusions of sodium and silicon. The latter is presumably present in the form of "trapped" gel. Both contaminants are non-uniformly distributed within the boundary region which is between 50 and 150μ thick (14), and the distributions are not coincident, as may be demonstrated by means of a scanning electron probe. After sufficient *in situ* etching, boundaries on regrown crystals are often free from such contamination. To make the *in situ* etching effective, the gel must be sufficiently acidic, e.g., at pH = 3. Experiments to determine how gel inclusions depend on gel density remain to be made. (See Section 3.4 for growth by hybrid methods.)

3.3 Cusp Formation

When crystals grow in primary gel media, they are often found associated with cusp-like cavities, i.e., regions in which the gel has been split and separated from the growing faces. These cusps may be large and obvious, but sometimes subtle lighting and a microscope are needed to reveal their existence. Those seen in the immediate vicinity of crystals arise, presumably, from the pressure of the advancing growth surface and are the outcome of gel displacement. They have been observed in a variety of growth systems; and while it is tempting to believe that they are always present in one form or another, the case of calcite discussed in Section 3.4 shows that exceptions are possible. (Examples of cusps are shown in Fig. 3.6.) There is, of course, no doubt that the cusps are filled with solution, but it had long been a matter of puzzlement as to why the crystals should grow just as well where the gel is in contact as they do where it is not. The contents of cusps have never been analyzed. For the sake of internal consistency, it is necessary to assume that the solute concentration within them equals that at the growth surfaces elsewhere on the crystal. Recent work by Hanoka (15) clarifies this situation. It will be seen, particularly from Fig. 3.6d, that neighboring cusps are in fact connected. As a result, the crystal is almost entirely surrounded, not by gel, but by solution; and this makes the equal-concentration hypothesis much more plausible. The present view, therefore, is that the crystals *nucleate* in the gel, but they grow increasingly from solution (except at a few points of support where, presumably, they rest upon the gel itself). The diffusion process supplies solute to the cusps and, in this way, governs the growth rate. It does not appear likely that convection currents within the small cusp volumes play any appreciable part in the proceedings.

The above comments apply to primary growths in which the development of crystal and cusps goes hand-in-hand. Reimplantation, followed by immediate growth, disturbs these relationships, since it causes crystals of substantial size to be confronted by initially cusp-less gel media. In these cases, the growth must be via diffusion through the gel until sufficient growth has taken place to cause cusps to be produced. The initial period after re-implantation is therefore similar to the initial period of primary growth, during which the chances of incorporating gel matter in the growing crystal are greatest. This is in agreement with the re-implanta-

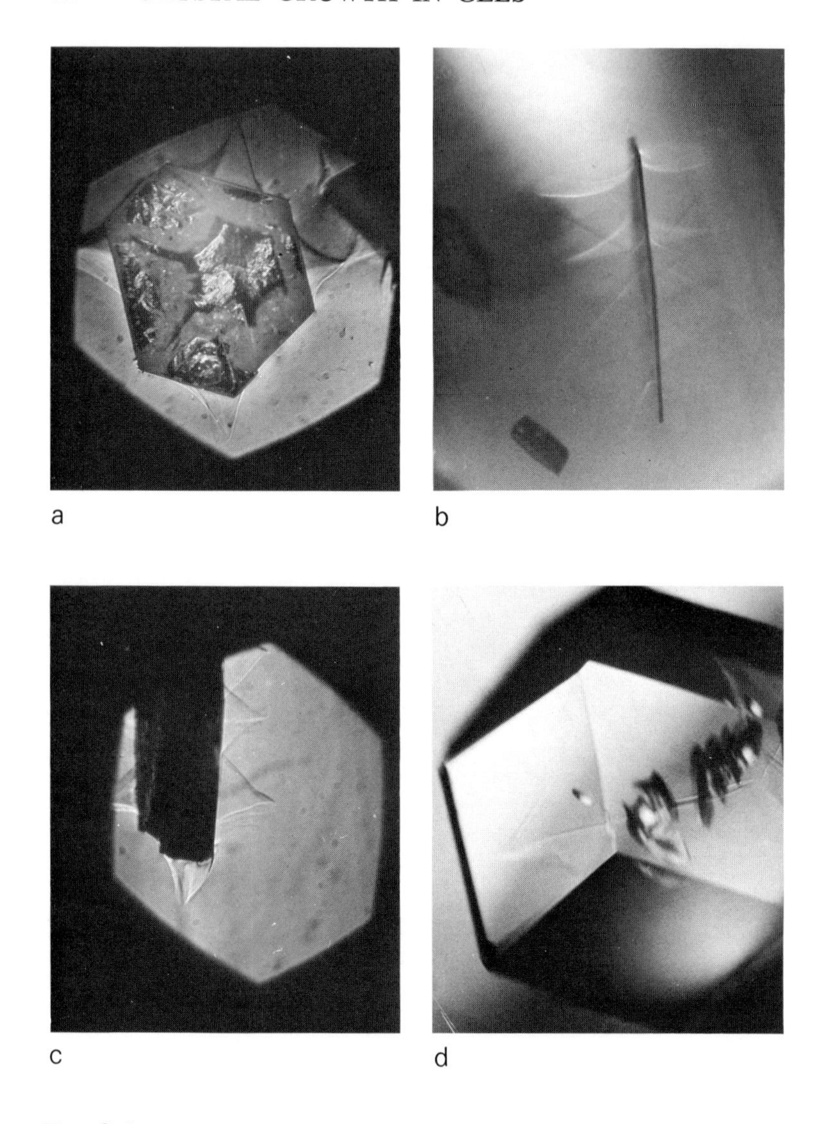

Fig. 3.6.

Cusp formation: (a, b and c) lead iodide, (d) calcium tartrate [After Hanoka, personal communication].

tion experiments described above and may also offer an alternative explanation for the existence of growth veils (compare Section 3.1). A crystal re-implanted and etched *in situ* for a sufficiently long time before growth does not suffer from this, since its initial regrowth is from solution, with only minimum gel contacts involved (14).

Subsequent growth differs from ordinary solution growth in some ways which are believed to be important. The surrounding gel permits diffusion which tends to replenish matter taken from the cusp volume by the growing crystal. It also protects the growth region from secondary (foreign) nuclei. Moreover, as long as the cusp volume is small enough, the solute concentration within it is self-regulating in the same sense as the surface-supersaturation discussed in Section 3.1.

3.4 Hybrid Procedures; Calcite Growth

Except for cusp formation as discussed above, all crystals grown by the basic techniques (Section 1.3) are completely surrounded by gel at least initially. However, whereas the gel is a good nucleation medium for many types of crystals, it does not follow that it is necessarily as good for subsequent growth. In the ordinary way, the uptake of silica by growing crystals is so small as to be quite unimportant, but in the case of calcite, for instance, the situation is very different.

Calcite crystals have well-known applications in optical instrumentation and laser technology; and since the sources of natural specimens appear to be diminishing, a special interest is attached to all methods of growing the material artificially. Previous attempts to grow calcium carbonate by a hydrothermal method have been described by Ikornikova and co-workers (16), experiments on growth in solution by Gruzensky (17) and Kaspar (18), and in electrochemical systems by Bárta and Žemlička (19). Morse and Donnay (20) and McCauley (21) have described growth in gels, the last with emphasis on reaction mechanism and phase aspects. The subject is also of interest in connection with the formation of calcium carbonate deposits in the human body (22).

The formation of calcite in Na-metasilicate gels is accomplished by the reaction between carbonates and Ca-salts (23, 24). Two methods have been developed for this purpose (25). In the first the gel itself contains the carbonate. An aqueous mixture of Na-metasilicate and a carbonate is prepared, and the pH is adjusted to between 7 and 8, usually by means of acetic acid. After the gel has set, a Ca-salt solution is put on top and allowed to diffuse. Attempts to mix the Ca-salt with the gel fail because Ca-silicate precipitates at and above a pH of 7. At lower pH values this precipitation is avoided, but there is a danger of CO_2 production which can destroy the gel. In the second method the neutral gel is initially free of calcium and carbonate ions. The reagents diffuse into it from two sides and form calcite where they meet. This is conveniently done in U-tubes or in tubes with fritted disk inserts (Fig. 3.7).

There appears to be no significant difference between the merits of the two methods; both produce well-shaped calcite rhombohedra of up to 5 mm size within 6 to 10 weeks. A few spherulites of aragonite and vaterite also grow. The three modifications have been verified by comparison of their *d*-values with those compiled

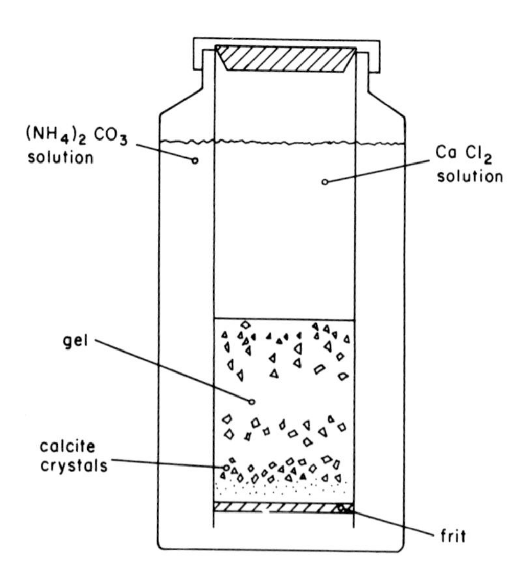

Fig. 3.7.

Calcite growth in tubes with fritted disks [After Nickl and Henisch (25)].

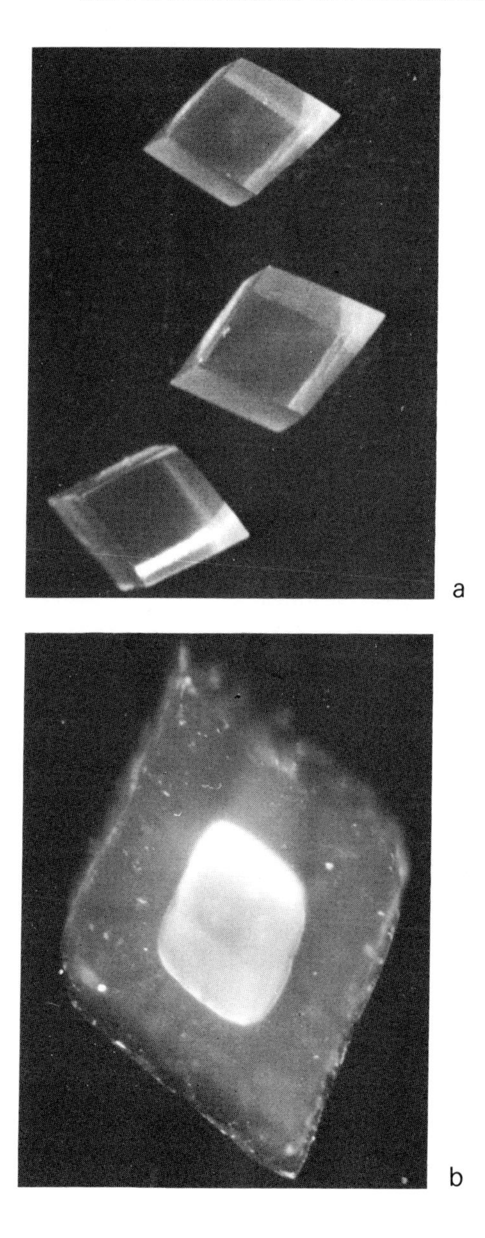

Fig. 3.8.

Gel-grown calcite crystals: (a) original form (3.5 mm maximum linear size), (b) after partial solution in acid [After Nickl and Henisch (25)].

by Swanson and Fuyat (26) and McConnell (27). Room temperature (25°C) appears to be optimum for growth. High temperatures, e.g., 70°C or so, favor the formation of aragonite (28, 29) and cause bubbles to form which disrupt the gel medium. This should be made from gelling solutions of analytic grade $Na_2SiO_3 \cdot 9H_2O$ with concentrations between 0.17 and 0.23 M. As a source of carbonate ions, solutions of Na_2Co_3, (pH 11.6); $(NH_4)_2CO_3$, (pH 9); $NaHCO_3$, (pH 8.6); and NH_4HCO_3, (pH 8.4) may be used; and the calcium can be conveniently derived from $CaCl_2$ or $CaAc_2$. The combination of $(NH_4)_2CO_3$ and $CaCl_2$ in equal concentrations (0.16 M) has been found to give the best results. Na_2CO_3 is less suitable on account of its high pH, which, upon neutralization, leads to high concentrations of sodium acetate. This, in turn, affects the quality of the resulting crystals adversely, whereas NH_4-acetate appears to have no such effect.

Crystals grown in the normal way as described are well-formed rhombohedra (Fig. 3.8a) but are almost invariably turbid, obviously due to inclusions. Spectroscopic and electron microprobe analyses have shown that they contain between 10 and 100 ppm of Mg and, much more important, between 0.45 and 1.7% of SiO_2, when prepared in gels of varying density between 1.02 and 1.03 g/cc. Despite this gross contamination, the specimens have well-developed and smooth crystal faces. Dissolution of a turbid crystal in acid leaves a residue which maintains the shape of the original specimen. Figure 3.8b shows this phenomenon. The dense inner core is part of the original calcite crystal. The surrounding residue turns out to be silica gel which can be examined by a freeze-drying technique (30) described in Section 4.5 and can be shown to have the same structure as the original growth medium (Fig. 3.9). The turbid crystals contain no water. It is clear from the results that the silica network which constitutes the gel is incorporated into the growing crystals more or less intact. In this way, calcite differs greatly from other gel-grown crystals (e.g., calcium tartrate), which are surprisingly free from silica contamination. In these cases, the gel is bodily displaced by the advancing growth surface, whereas calcite permeates the silica network while maintaining a high level of short-range order. In this respect, the gel-grown specimens resemble certain natural calcite structures, e.g., the spikes of sea urchins.

A few of the crystals are found to grow in fissures, and thus at the boundaries between gel and liquid. These are turbid to the

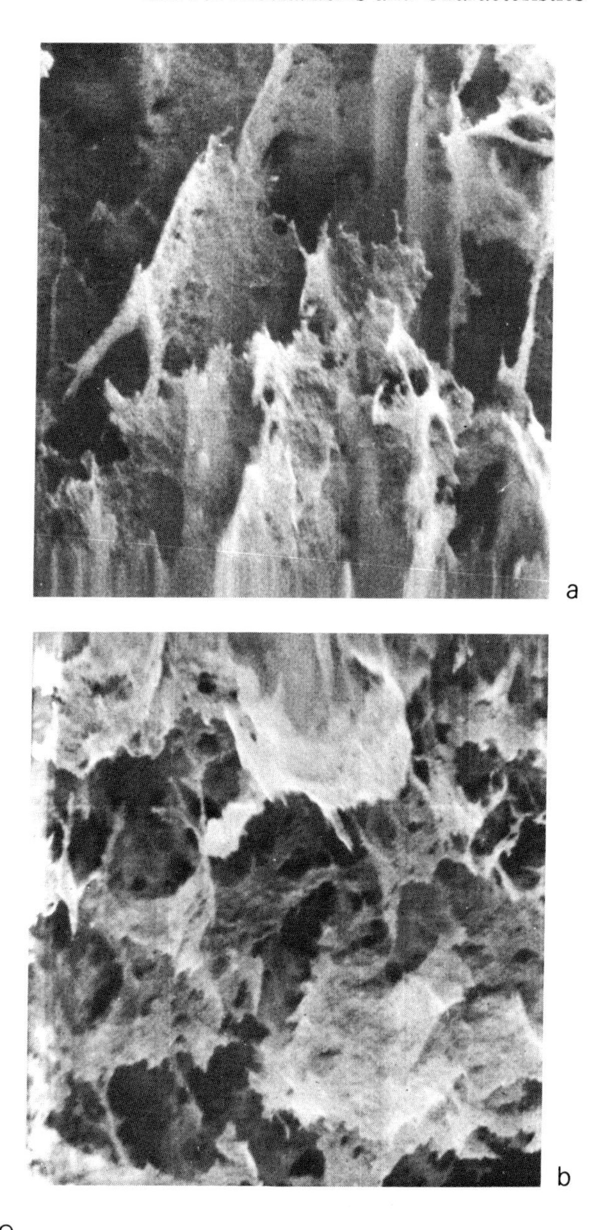

Fig. 3.9.

Silica gel structures revealed by a scanning electron microscope after vacuum freeze drying; x 2300: (a) residue after crystal dissolution, (b) normal growth medium [After Nickl and Henisch (25)].

extent to which they overlap the gel and clear to the extent to which they grow in solution. The two regions can be clearly seen (Fig. 3.10). The same observations have been made on crystals of aragonite and vaterite grown by these procedures. This suggests that calcite and other crystals which may be found to behave in a similar way should not be grown in gels at all. They should be grown in solution, if possible without losing the general benefits of the gel method. The solution should thus be in a cavity of very small volume, so that the solute concentrations within it may be regulated by the demands of the growing crystal. To ensure this, the size of the cavity should always be comparable with the volume of the growing crystal, a condition not easily achieved, and especially not during the initial stages of growth. The solute must be replenished by diffusion in order to achieve the self-adjusting boundary concentration noted in Section 3.1. The diffusion medium must be a gel, so as to prevent secondary (foreign) nuclei from reaching the vicinity of the growing crystal (see Chapter 4). These requirements distinguish the optimum arrangement from the systems, otherwise similar, proposed by Torgesen and Peiser (31). The solution-filled growth cavity could be seeded with a small calcite crystallite which may originate from a pure gel system but need not do so, especially if clear specimens are available from other sources.

Hybrid procedures of this kind are under investigation but have not yet been widely explored. Figure 3.11 gives two examples which have proved successful, even though they manifestly fail to satisfy the small volume requirement. In the test-tube system on the left, sodium metasilicate solution is floated upon a highly concentrated solution of NH_4Cl and allowed to gel before adding the supernatant $CaCl_2$. Crystals then grow mostly in the solution belt. Unwanted nucleation on the wall of the tube can be diminished by giving it an initial gel coating. The system on the right involves diffusion of the reagents through separate gel columns, and the growth medium itself should also be pre-filtered in this way. Epitaxial growth on the seed occurs, and the new layers are clear, no matter whether the seed contains a silica network or not. When clear seeds are used, no boundary between substrate and new growth can be detected. Weight increases by factors up to 10 have been recorded (25). As far as is known, crystals large enough for optical applications have not yet been grown, but there is a real possibility that the gel method might yet come to be used for

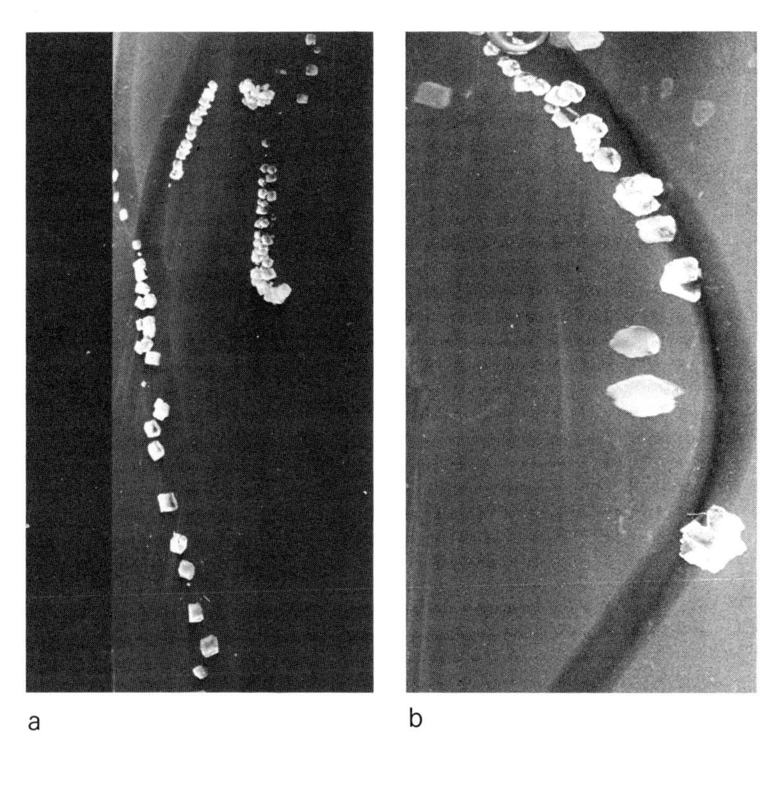

Fig. 3.10.

Calcite crystals with growth boundaries: (a) and (b) crystals growing partly in solution, (c) boundaries of gel inclusion [After Nickl and Henisch (25)].

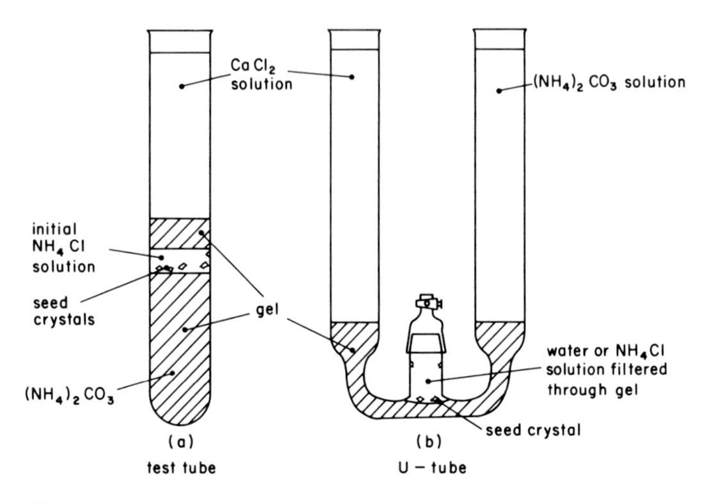

Fig. 3.11.

Systems for calcite growth by hybrid gel methods [After Nickl and Henisch (25)].

this purpose. A mathematical analysis of schematic diffusions systems, with particular reference to the local supersaturations produced, has been provided by Lendvay (32).

The methods described above can also be used for the preparation of doped crystals. Dopants in the form of chlorides or nitrates can be incorporated, before gelling, in the sodium metasilicate solution or in the supernatant solution. Small amounts do not appear to affect the growth habit but high dopant concentrations prevent the growth of regular rhombohedra. The maximum permissible concentrations for regular growth differ for various metal ions (25). For Co^{+2}, Ni^{+2}, Cu^{+2}, Mn^{+2}, Mg^{+2}, and Zn^{+2}, they are of the order of 10^{-3} M in the gel and 5 to 20 times higher in the supernatant solution. For Cr^{+3} the maximum is about 0.1 M in the gel. As it happens, all these metals form hydroxides which are insoluble in pure water but somewhat soluble in the presence of NH_4OH and NH_4-salts. For Nd^{+3}, Ho^{+3}, Er^{+3}, and Fe^{+3}, the maximum permissible concentrations in the gel are of the order of 10^{-4} M or, as in the last case, are too small to be readily determined. These metals form hydroxides and carbonates which are even more insoluble (under the growth conditions employed) than those listed above. It is therefore reasonable to assume that they precipitate earlier in microcrystalline form. Then they would constitute heterogeneous nuclei (see Section 4.3) on which calcium carbonate growth could take place epitaxially, as suggested by McCauley (21). Epitaxy would lead to growth forms which have no particular relation to the rhombohedra found at lower dopant concentrations. For the metals which form more soluble hydroxides and carbonates, the same effect would occur but only at higher concentrations. This interpretation is in harmony with the fact that dopant concentrations which are high enough to destroy regular growth also lead to greatly increased nucleation. Valency considerations must, of course, enter into this picture, but the high permissible concentrations of Cr^{+3} discount their dominating importance.

There is an enrichment of dopants in the crystals by a factor of about 100, as compared with the average dopant concentration in the gel. This probably arises from the fact that the dopant establishes its own radial concentration gradient around the growing crystal. The volume from which dopant is drawn is thus likely to be many times larger than that occupied by the growth itself. Electron microprobe tests have shown (25) that the dopants are uniformly distributed within the crystals or, at any rate, unclustered, whether these are clouded by silica networks or not. Of the crystals doped with the above elements, only those containing Mn were found to be photoluminescent (orange) and cathodoluminescent (red).

References: Chapter 3

1. Vand, V.; Vedam, K.; and Stein, R. *J. Appl. Phys.* 37:2551 (1966).
2. Frank, F. C. *Proc. Roy. Soc.* 201A:586 (1950).
3. Sultan, F. S. A. *Phil. Mag.* 43:1099 (1952).
4. Henisch, H. K.; Hanoka, J. I.; and Dennis, J. *J. Electrochem. Soc.* 112: 627 (1965).
5. Dennis, J., and Henisch, H. K. *J. Electrochem. Soc.* 114:263 (1967).
6. Dennis, J. *Crystal Growth in Gels.* Pennsylvania State University Thesis (1967).
7. Faust, J. W. Pennsylvania State University, personal communication (1968).
8. Egli, P. H., and Johnson, L. R. In *The Art and Science of Growing Crystals* (Ed. J. J. Gilman). John Wiley & Sons, New York (1963).
9. Armington, A. F.; O'Connor, J. J.; and DiPietro, M. A. Air Force Cambridge Research Laboratories (Reference 67–0304), Physical Sciences Research Paper No. 325 (May, 1967).

10. Jagodzinski, H. *Crystallography and Crystal Perfection.* (Ed. G. N. Ramachandran). p. 177. Academic Press, New York (1963).

11. Washburn, J. *Growth and Perfection of Crystals.* (Eds. R. H. Dorman, B. W. Roberts, and D. Turnbull). p. 342. John Wiley & Sons, Inc., New York (1958).

12. deHaas, Y. M. *Nature.* 200:876 (1963).

13. Khaimov-Malkov, V. Ia. *Soviet Physics, Crystallography.* 2:487 (1957).

14. Bulger, G. Pennsylvania State University, personal communication (1969).

15. Hanoka, J. I. *J. Appl. Phys.* 40:2694 (1969).

16. Ikornikova, N. Yu, and Butuzov, V. P. *Doklady Akad. Nauk SSSR.* 111:105 (1956).

17. Gruzensky, P. M. *J. Phys. Chem. Solids.* Supplement No. 1. p. 365 (1967).

18. Kaspar, J. *Growth of Crystals.* 57:2 [Interim reports between the first (1956) and second Conference on Crystal Growth, Institute of Crystallography, Academy of Sciences, USSR.], Consultants Bureau, New York (1959).

19. Bárta, C., and Žemlička, J. Institute of Solid State Physics, Prague, personal communication (1967).

20. Morse, MM. H., and Donnay, D. H. *Bull. Soc. Franc. Mineral.* 54:19 (1931).

21. McCauley, J. W. M.S. Thesis in Geochemistry and Mineralogy, Pennsylvania State University (June, 1965).

22. Beck, C. W., and Bender, M. J. *J. of Urology.* 101:208 (1969). See also Pfeiffer, R. R. *Research Today.* 15:2 (1959). Eli Lilly Co., Indianapolis.

23. Hatschek, E. *Kolloid Zeitschrift* 8:13 (1911).

24. Fisher, L. W., and Simons, F. L. *Amer. Mineralogist.* 11:124 (1926).

25. Nickl, J., and Henisch, H. K. *J. Electrochem. Soc.* 116:1258 (1969).

26. Swanson, H. E., and Fuyat, R. K. *NBS Circular* 539, II:51 (1953) and III:54 (1953).

27. McConnell, J. D. C. *Min. Mag.* 32:535 (1960).

28. Kitano, Y. *Bull. Chem. Soc. Japan.* 35:1980 (1962).

29. Dekeyser, W. L., and Degueldre, L. *Bull. Soc. Chim. Belg.* 49:40 (1950).

30. Halberstadt, E. S.; Henisch, H. K.; Nickl, J.; and White, E. W. *J. Colloid & Interface Science.* 24:461 (1969).

31. Torgesen, J. L., and Peiser, H. S. *Methods and Apparatus for Growing Single Crystals of Slightly Soluble Substances,* U. S. Patent 3,371,038, February 27 (1968).

32. Lendvay, O. *Magyar Fiz. Fol.* 8:231 (1965). In Hungarian. English version: Air Force Cambridge Research Laboratories, Translation No. 51 (Ref 69–0275), June, 1969.

4 Nucleation

4.1 General Principles

The problem of nucleation is of crucial importance in practical operations, since the crystals which grow in any particular gel system compete with one another for solute. This competition limits their size and perfection, and it is obviously desirable to suppress nucleation until, ideally, only one crystal grows in a predetermined location. The available techniques have not yet reached this level of perfection, though they can sometimes approach it.

Since the application of dislocation theory to these problems by Frank and co-workers (1, 2, 3), there has been a great increase in our knowledge of the manner in which crystals continue to grow, once growth has started. In comparison, the amount of precise information on the nature of that start is still only small. Always experimentally difficult, the problem is evidently simplest in vapors and melts because only one substance is involved in such systems. It is *a priori* more complex in the case of solutions because of solute-solvent surface interaction and the possibility of nuclei in the course of formation being solvent contaminated (4).

Crystal growth in gels is evidently a variant of growth in solution, with additional complications arising from the presence of the gel. In this sense, gel systems do not lend themselves well to nucleation studies of the most fundamental kind. Detailed quantitative considerations, though superficially tempting, are therefore not (or, at any rate, not yet) likely to be profitable in the present context. On the other hand, it is now abundantly clear that gels reduce the nucleation probability and, in this sense, offer certain research opportunities which experiments on growth in liquids cannot provide. The nucleation suppressing character (see below) distinguishes gel methods from ordinary diffusion methods sometimes used for crystal growth (e.g., 5).

It has long been known that the formation of crystals is sensitively dependent on the presence of impurities. The first systematic and quantitative investigations of this kind were carried out by Tammann (6), mainly on crystallization from the melt. He found that crystallization could be increased by soluble, as well as insoluble (e.g., quartz powder), additives. Since then, two basic nucleation mechanisms have been recognized: *homogeneous nucleation,* which does not fundamentally call for the presence of foreign substances (even though it can be influenced by them); and *heterogeneous nucleation,* which demands a preexisting foreign crystalline substrate* on which new material can be deposited from the vapor, the melt, or from solute. One is then dealing with a form of epitaxial crystal growth. It is believed, for instance, that silver iodide serves as an advantageous foreign nucleus during cloud seeding because its lattice constant is close to that of ice (7). Similarly, microcrystals of sodium sulfate are excellent crystallization substrates for sodium carbonate, and phosphates may be used to induce crystallization in solutions of arsenates (8). The extent (if any) to which noncrystalline particles can serve as foreign nuclei is unknown.

In practice, it is impossible to free systems entirely from foreign particle contamination, and this makes it impossible to guarantee that all the observed nucleation is ever homogeneous. On the other hand, it is feasible (see below) to devise experiments which demonstrate that both types of nucleation can occur and, in particular, that homogeneous nucleation sometimes prevails in gels. Of the two processes, the deposition of solute on a pre-existing substrate is energetically the "cheaper." It therefore occurs at lower supersaturations and one must expect homogeneous nucleation to be delayed until the available heterogeneous nuclei are used up. Because epitaxial growth as such lends itself to experimentation on a macroscopic scale, its mechanisms are better understood. Even then, there is still uncertainty as to the degree of lattice matching between substrate and deposit required for single crystal formation. Moreover, the presence of foreign atoms can affect the mutual binding energy and thus influence the heterogeneous nucleation probability. A number of workers have concerned themselves with the manner in which different crystalline substances affect crystallization from solution (e.g., see 7, 9, 10).

* This being the term with which we dignify the scientifically tractable part of what is otherwise known as "dust."

The range of possibilities is obviously large, from complete lattice mis-match, on the one hand, to a substrate which is isomorphic with the growing crystal, on the other. A review of "catalyzed nucleation," as the process is also called, has been given by Turnbull and Vonnegut (11), although a great deal of work has been done since then (1952).

All theories of homogeneous nucleation involve the concept of the critical nucleus. It is envisaged that, as a result of statistical accident, a number of atoms, ions, or molecules can come together and form a rudimentary crystal. Simple energetic considerations show that this crystal is likely to dissolve again unless it reaches a certain critical (minimum) size. Beyond that size, the energy relations favor continued growth. This does not necessarily mean that a macroscopic crystal will grow, since this would depend on the availability of sufficient solute. It does mean, however, that the assembly is stable under the prevailing conditions. If the conditions change, e.g., by the appearance of new concentration gradients due to the growth of other crystals in the neighborhood, the growth rate will adjust itself accordingly and may even become negative. It is a common observation that large crystals grow at the expense of smaller ones, and some of these may entirely disappear. The physical reality of the critical nucleus was first demonstrated by Ostwald (12), and many attempts have since been made to determine typical sizes. However, because critical nuclei are so small, accurate measurements are not really feasible, and calculated values depend very much on the assumptions made, many of which cannot as yet be independently tested. Calculations (13) based on the simplest (though not the most plausible) models lead to critical radii of the order of 10Å, much too small to be seen even by the electron microscope. The inferences drawn from observations are therefore always indirect.

It is common practice (8) in the analysis of nucleation phenomena in solutions to rely heavily on analogies with nucleation in the vapor phase. As is well known, small droplets (radius r) of liquid have higher vapor pressure (p_r) than a flat liquid surface, a fact expressed by the so-called Gibbs-Thomson formula, which can be written in the form

$$\log \frac{p_r}{p_\infty} = \frac{2M\sigma}{RT\rho r}. \tag{4.1}$$

In this equation, M is the molecular weight, σ the surface energy per unit area (here assumed independent of r), R the gas constant, T the temperature and ρ the droplet density. All the analogies depend on the notion that p_r/p_∞ can be simply replaced in solution by the macroscopic supersaturation, the saturation being defined by reference to the solute concentration, C_∞, at which particles of infinite radius could nucleate. The nucleation of smaller particles would demand higher concentrations C_r in accordance with the equation

$$\log \frac{C_r}{C_\infty} = \log S_r = \frac{2M\sigma}{RT\rho r}, \qquad (4.2)$$

which also gives the required minimum value of the macroscopic supersaturation S. It has long been known that small solid particles are more soluble than large ones, a fact first analyzed by Ostwald in 1900 and much tested and elaborated ever since (e.g., 14–17). In that sense the analogy with the vapor pressures of small droplets is valid. It was recognized at an early stage that difficulties would arise if the particle radius r were allowed to go to zero. To avoid this problem, Knapp (18) postulated that the particles carry a small electric charge which can be shown to reverse the trend of the supersaturation at very small values of r. Further corrections were introduced (17) by envisaging the dissociation of the solute and other factors. However, whereas equation 4.2 and its varied elaborations may be a reasonable approximation of the facts, they are not by themselves helpful in gaining further insight. It is therefore necessary to limit the argument to more general considerations. The classical argument is that the establishment of a homogeneous spherical nucleus of radius r *releases* an amount of energy equal to $4\pi r^3 \rho L/3$, where L is the heat of transition involved (in this case, the heat of solution). On the other hand, the establishment of a surface demands energy to the extent of $4\pi r^2 \sigma$. Other geometries have been analyzed, but as long as there is no definite knowledge of the manner in which σ varies for different crystal surfaces such geometrical refinements are not profitable. The total energy change is evidently a function of r of the form shown in Fig. 4.1, in which the position of the maximum defines the critical radius r_c, such that

$$r_c = \frac{2\sigma}{\rho L}. \qquad (4.3)$$

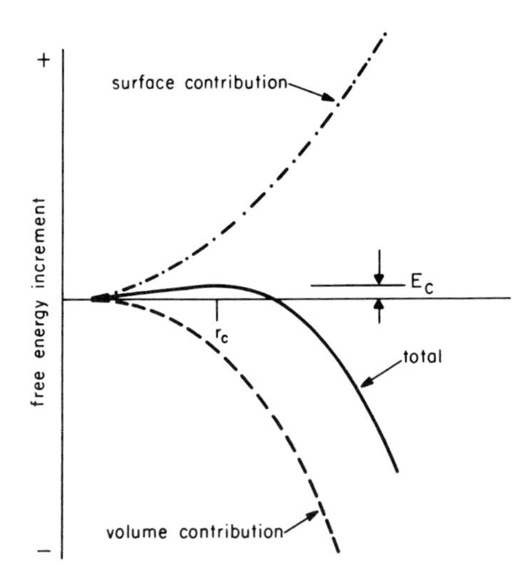

Fig. 4.1.

Role of volume and surface energies in the determination of the critical nucleus radius r_c. Classical picture; perfect spherical nucleus assumed.

The energy which has to be available for the creation of this nucleus becomes

$$E_c = \frac{16\pi\sigma^3}{3\rho^2L^2}, \tag{4.4}$$

and the nucleation probability should itself be proportional to $\exp(-E_c/RT)$. Although the average volume of E_c must be taken as constant at a given temperature and pressure, local and temporary fluctuations from this average are believed to be possible. At points characterized by temporary minima, nuclei are most likely to form.

This is the classical model, and that it constitutes a gross oversimplification of the problem is obvious enough. However, whereas its quantitative aspects must not be taken too seriously, it remains useful as a basis of qualitative discussions. It is easy to see, for instance, that there should be an optimum temperature for nucleation. As the temperature increases, the probability of E_c being available increases but the degree of supersaturation for particles of critical nucleus size diminishes, in accordance with the appropriate form of equation 4.2:

$$\log \left(\frac{C_r}{C_\infty}\right) = \log S_c = \frac{ML}{RT}. \quad (4.5)$$

That the nucleation rate goes through a maximum at a definite temperature was first confirmed by Tammann (6), not indeed for solutions but for supercooled melts. Köppen (19) many years later provided similar evidence for KCl solutions. At all temperatures, some of the nucleation is delayed in accordance with statistical factors, and additional delays may occur in viscous media or in systems in which there are other transport problems, e.g., in gels. Under dynamic conditions, this can lead to substantial build-ups of supersaturation.

Models which postulate spherical nuclei cannot, of course, account for the existence of crystals in different growth habits. In some cases, the factors which favor a particular habit are known. Thus, when a great deal of heat of crystallization has to be dissipated, dendrite and needle growth are obviously advantageous, because of the large surface-to-volume ratio. In most other cases, the origin of habit-determining factors is obscure. The ignorance is general, and is not confined by crystal growth in gels; many empirical observations have been reviewed by Buckley (20). Although the crystal habit is evidently determined during the early stages of growth, there is some evidence (e.g., see Section 4.2) that the "habit code" is not irrevocably governed by the structure of the nucleus. It can also be influenced by later circumstances.

The effect of macroscopic supersaturation, as such, can be more directly seen by writing equation 4.3 in an alternative form, namely (with σ in appropriate units)

$$r_c = \frac{2\sigma\nu}{kT \log S}, \quad (4.6)$$

in which ν is the molecular volume and k, Boltzmann's constant. This is derived (13) by considering the volume contribution to the free energy to be made up of the chemical potential μ of each particle referred to the solution. From this and $\mu = kT \log S$, equation 4.6 follows directly. In the same terms, the critical energy becomes

$$E_c = \frac{16\pi\sigma^3\nu^2}{3\ (kT \log S)^2}. \quad (4.7)$$

[See also Becker (21, 22) and Smakula (4)]. Increasing supersaturation thus increases the nucleation probability by diminishing E_c. At the same time, equation 4.7 suggests the possibility of experiments with solutes of different molecular volume. It has been amply tested for condensing vapors, and more sophisticated versions of it have been developed for crystallization from solutions, e.g., to take account of the effect of concentration on the chemical potential (23). The interface energy σ depends, of course, on the solvent as well as the solute, a fact demonstrated in the course of classic experiments by Amsler and Scherer (24, 25). They crystallized KCl from aqueous solutions which were progressively modified by the addition of alcohol. A corresponding shift in the nucleation probability was duly observed. Kamenetzkaja (26) has shown that the effect of impurities on σ is very complex, inasmuch as σ increases in some cases and decreases in others. Moreover, the effect which impurities have on nucleation may be very different (27) from their effect on growth. Apart from the studies described in the following section, there is no record of previous experiments concerned specifically with nucleation in gels.

4.2 Evidence for Homogeneous Nucleation

In any reasonably well-optimized gel system, only a few crystals grow to macroscopic size, and the identification of their nucleation mechanism is not a simple matter. That heterogeneous nucleation is possible is no longer in doubt. The phenomenon has been demonstrated by the deliberate addition of foreign nuclei (28) and, in the case of certain (but not all) PbI_2 crystals (see Section 4.3), by the detection of Ag at the growth center (silver being present in the reagents as a contaminant). The problem is to ascertain whether such nucleation as remains under cleanest conditions is likewise heterogeneous and thus dependent on very small, possibly subanalytical, foreign nuclei or homogeneous and dependent on the formation of a native nucleus of critical size, as governed by thermodynamic considerations. Because both processes are expected to depend on solute concentration, pH, impurities, and molecular size, incontrovertible proof is not available, but the following observations (29) collectively support the hypothesis that homogeneous nucleation can and does occur:

(a) For the growth of calcium tartrate, it is customary to diffuse $CaCl_2$ from a solution into the silica gel charged with tartaric acid, as described (Section 1.3). However, it is possible to pre-supersaturate the growth medium with calcium tartrate by adding a small amount of $CaCl_2$ before the gel sets. Such a gel is clear to the naked eye, but when examined under the microscope between crossed Nicols, numerous crystallites of calcium tartrate are clearly visible. As a result of subsequent $CaCl_2$ diffusion from the supernatant liquid, a very large number of small crystals appear in due course, uniformly distributed over the whole volume. If it were true that these are nucleated by foreign particles, the experiment would show that such particles are uniformly distributed in the gel as, indeed, one would expect. When the gel is not pre-saturated in this way, the usual highly nonuniform distribution of crystals develops in the course of diffusion, the number per unit tube length diminishing with increasing distance from the gel-solution interface. In many instances a good crystal eventually forms in some of the lower regions of the tube, showing that there is sufficient solute concentration, even though an inch or more of tube above it is entirely free from visible signs of nucleation. Since the concentration of calcium tartrate is bound to diminish with increasing distance from the gel interface, and since heterogeneous nucleation should demand lower supersaturations than homogeneous nucleation, these observations cannot be consistently explained on the assumption that they depend on foreign nuclei. Instead, we must assume that homogeneous nuclei of critical size can ordinarily form only in a few locations of the gel—and even there only in the presence of high supersaturations.

A complication arises from the fact that the concentration gradient is not everywhere parallel to the tube axis, but has radial components in the immediate vicinity of each new crystal. As the diffusion regions interact, the pattern of gradients ceases to be simple. It depends then on the location of the crystals formed and on the time sequence of their appearance.

(b) When PbI_2 crystals are examined during the initial stages of growth in a gel, they are often (if not always) found to be free from optically detectable defects. Such defects arise during later stages of growth, but it is hard to reconcile the geometrical and structural perfection of the smallest and thinnest crystals with any hypothesis of growth on a foreign substrate located (as it would have to be) at the center of the PbI_2 hexagons. These crystals are very thin but depend, presumably, on the initial formation of a three-dimensional nucleus. Once a rudimentary sheet structure exists, its thickness might be increased somewhat through the agency of two-dimensional surface nuclei as described by Mott (30), even before screw dislocations are formed, which account for all major growth in the c-direction. This can be done through the agency of dirt particles coming into contact with previously perfect growth surfaces. It is believed that the particles need not be crystalline to be effective in this sense.

(c) It is a common observation that many different crystals can be grown in a gel, some nucleating very readily and some with difficulty. (Calcium tartrate is a good demonstration crystal precisely because it nucleates only sparingly.) The differences are great, and it is hard to see why foreign nuclei should be so discriminating in their influence on nucleation frequency, while yet supporting in some degree the nucleation of a wide variety of substances. To save the heterogeneous nucleation hypothesis, it would be necessary to postulate that *different* nuclei are effective for different substances and that these nuclei are themselves encountered with different though reproducible frequencies in different gels. At that stage the hypothesis becomes too complicated and artificial to be convincing.

(d) Some gel growth experiments are remarkably insensitive to the filtering procedures employed. Although at least some of the foreign particles which are capable of acting as nuclei are presumably too small to be efficiently removed by filtering, some ought to be large enough, and it is thus difficult to reconcile the observations in a general way with any theory which ascribes a predominating function to such nuclei. (See, however, Section 4.3.)

(e) Experiments by Kratochvil and co-workers (31) on the gel growth of metallic gold crystals (from gold chloride, reduced by oxalic acid) have yielded triangular hexagon- and needle-shaped platelets, occasionally with hexagonal pyramids on the growth faces. These configurations (Fig. 4.2) bear a striking similarity to *vacancy* clusters observed in quenched gold prepared by other means. The two structures cannot plausibly be related, except through some process of homogeneous nucleation.

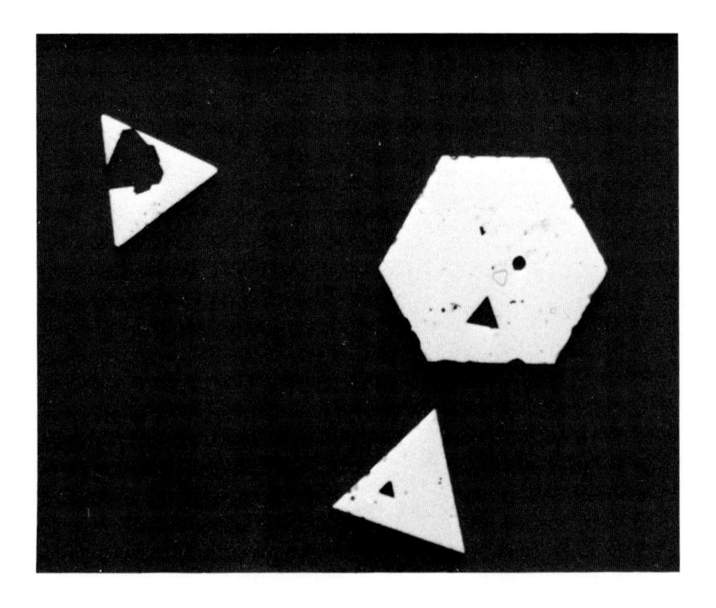

Fig. 4.2.

Photomicrographs of metallic gold crystallites [After Kratochvil, Sprusil and Heyrovsky (31)].

4.3 Studies on Heterogeneous Nuclei; Filtering

Because foreign nuclei are so small, their unequivocal detection and identification *in situ* is not an easy matter. It has only once been achieved for crystals grown in gels, in connection with an investigation by Hanoka (32, 33) on the origin of polytypism in PbI_2. (See also Section 5.1.) In some of the growth systems, a mass of small crystals were found to have formed near the bottom of otherwise only sparsely populated tubes. This suggested that the small crystals might have grown on foreign nuclei heavy enough to have settled before the gelling process or during its initial stages. To do this, the nuclei would have had to be of a certain minimum size. Estimates indicated that some, at least, should have been large enough to be visible under the microscope and this was confirmed by observations. Some of the PbI_2 platelets had visible spots of sizes up to 100μ, roughly at their geometrical centers (Fig. 4.3). They were at first believed to be on the surface

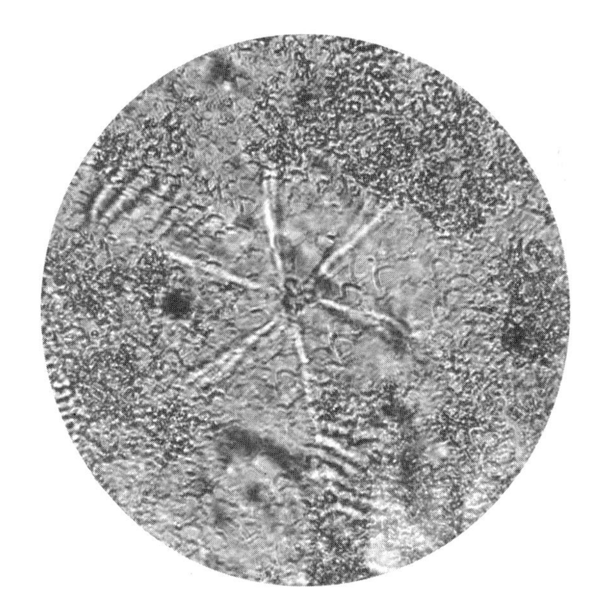

Fig. 4.3.

Lead iodide crystals containing a foreign nucleus. Field of view: 0.5 mm diameter [After Hanoka (32)].

but were in fact in the interiors of the crystals. A slicing technique had to be applied to make them directly accessible for identification by a scanning electron microprobe. The fact that PbI_2 is cathodoluminescent permitted the centers to be quickly located under the electron beam, since the emission spectrum was noticeably different at these places. The presence of silver was verified by reference to the L_α and L_β lines and their comparison with a silver standard. It had its origin in the "reagent grade" lead acetate which permeated the gels, in which it occurred as a notably unlisted but nonetheless prominent contamination.

The presence of the silver nuclei was found to disturb the crystalline order in a variety of ways. In some cases, it gave rise to six radial lines of defects emanating from the center and likewise contaminated with silver. In other cases the silver caused voids, and in some the foreign nuclei were seen to be located at the center of spirals or at the common center of a series of hexagons. These findings have an important bearing on the theory of polytypism. They cannot prove that *all* polytypism arises from screw dislocations, but they make it highly plausible that at least some forms of polytypism do, as originally suggested by Frank (34, 35) and Vand (36).

In most gel systems there is some growth on the surfaces of the container (usually glass), and experiments on reimplantation (Section 3.2) yield evidence of the fact that boundaries between two gels solidified at different times are favored as nucleation sites. In a similar way, crystals often grow in contact with macroscopic bubbles in the gel or at internal gel surfaces created by mechanical rupture. All these cases must be regarded as instances of heterogeneous nucleation, but almost nothing is known about the processes involved. (See also Section 4.7.) To diminish unwanted nucleation arising from gas bubbles, the constituents of the gelling solution should be boiled before mixing.

In an attempt to gain further information about heterogeneous nuclei, it is necessary to explore the effect of foreign additives and the effect of filtering procedures. As far as crystalline additives are concerned, there is some evidence (37), though not of a highly formal kind, that they do increase the number of crystals formed. Controlled experiments are not simple because they depend on the chemical surface history of the dispersed powders. There is always a likelihood that small gas bubbles may adhere to the foreign crystallites and lead to misleading results. Soluble additives offer more reliable opportunities for observation. Among those tried, e.g., by Perison and co-workers (38), are various types of sugars. When sucrose, for instance, is incorporated in the gelling mixture, it leads to a sharp increase in nucleation probability. This has been demonstrated for calcium tartrate and AgI. Sucrose also affects the gel structure, as may be seen from the greatly increased transparency of the medium. This left open the possibility that the effect of sucrose might be simply "structural" (see Section 4.5) and not directly due to heterogeneous nucleation. To test this, sucrose was alternatively diffused with the $CaCl_2$ into an otherwise normal tartaric acid gel. A large increase in nucleation (e.g., by a factor of 10) and also a change of growth habit were observed, needles being preferred. Nucleation in AgI growth systems was also increased. Similar experiments have been made with dextrose. When present in the gel from the beginning, dextrose increased AgI nucleation by 12% and calcium tartrate nucleation by 100%. When the same dextrose concentrations were employed as supernatant diffusants, the corresponding increases were 412% and 110%. Gel structure effects may be at work, but the results suggest that certain organic molecules can themselves act as nucleation centers, an interesting possibility which deserves further exploration.

Experience has shown that the use of ordinary filter papers does not lead to any consistent reduction in the number of crystals grown. It has, indeed, been known at times to promote nucleation! The observed inconsistencies are no doubt connected with the fact that cellulose is somewhat soluble in sodium metasilicate solution. Proprietary ("Millipore Solvinert") filters give more consistent results. They too lead to increased nucleation in the case of calcium tartrate, but to a significant reduction (factors of about 3) in the number of AgI crystals grown. Filters of 1.5μ mean pore size have proved most effective (38) for the sodium metasilicate solution. Filtering of the KI solution incorporated in the gel (using 0.25μ mean pore size) produced a further, smaller nucleation drop. Most significantly, filtering of the supernatant-complexed solution had no effect at all. This shows convincingly that diffusion through the gel is itself the most effective filtering process, at any rate, as far as heterogeneous nuclei above a certain size are concerned.

Sugar molecules (see above) are evidently smaller. "Millipore" filters are thus effective but not completely so. Spectroscopic analysis of the filter plates has shown that at least some foreign iodides are indeed removed from the solution during filtering, but substances which manage to pass can act as nucleation centers when present within the pore structure of the gel. This can be demonstrated simply. When small amounts of AgI powder are suspended in the gel during its formation, they lead to an enormous increase in the number of crystals formed. Self-epitaxy is, of course, the most likely process, but there is no reason to believe that other iodides which are isomorphous with AgI (or nearly so) will behave differently. At the same time, the number of visible crystals formed is orders of magnitude smaller than one would expect from the number of powder grains added. The simplest explanation is probably the best: that a great deal of nucleation is suppressed by gel envelopment of foreign substrates, which prevents solute from reaching them in amounts sufficient for macroscopic growth.

The above results suggest that optimum conditions for nucleation suppression would prevail if all components of a growth system were filtered through gel. This is a simple matter as far as all the reagent solutions are concerned, and the success of such procedures is well demonstrated by the hybrid methods described in Section 3.4. However, sodium metasilicate solution itself does not easily pass through a gel, and a really satisfactory method of filtering it remains to be found.

4.4 Nucleation Control

The suppression of nucleation is the principal function of the gel, but it is apparent that the degree of suppression ordinarily obtained is insufficient for many of the crystals one wishes to grow. As it happens, calcium tartrate nucleates only sparingly; and that fact, rather than its compelling fascination, is why it is so often used for crystal growth demonstrations. It is, in fact, a remarkably unexciting material but, precisely because it grows so well, it lends itself to a variety of studies on the behavior of gel systems. Many other crystals nucleate more easily and, as a result, grow less well. In such cases, there is every incentive for a search to discover additional means of nucleation control.

In principle, the simplest way of achieving an additional reduction of nucleation would be to accentuate the "natural" suppressive action of the gel. This, in turn, must have something to do with the pore size distribution and with the extent to which pores are in communication with one another. The simplest visualization suggests that homogeneous nuclei of critical size cannot form in very small, isolated pores because the necessary amount of solute is not available. Critical nuclei may form in larger pores, but they are not expected to grow to macroscopic size unless there is a sufficient degree of communication with other pores containing solute. Much the same would apply to heterogeneous nuclei. Some of these might be completely "protected" by gel matter and thus be inoperative; others might happen to be in suitable locations in which continued growth can be supported by the prevailing diffusion condition. Some potential nuclei can be "used up" by the gel structure itself, held by chemical bonding forces in the three-dimensional silica network and thereby rendered ineffective as crystallization substrates. The above results show that this happens with sugars, but there is no evidence that insoluble crystalline additives are built into the gel network in this way.

The present picture ignores any part which may be played by the pore walls, but it suggests that there are favorable and unfavorable pore size distributions. The comparison between silica and gelatin gels (Section 4.5) bears this out in rough terms, and the model is also in agreement with experiments involving PbI_2 growth in gels of different density (39). However, far too little is known about gel structures to permit firm conclusions. In many cases, though lamentably not in all, greater gel density (which intuitively implies smaller pore sizes) does indeed diminish nucleation. On the other hand, it tends to increase the contamination of the crystals by silicon and, thereby, to spoil their perfection and shape. The density used in practice is therefore a compromise.

Another method of nucleation control involves the deliberate addition of foreign elements and is, of course, open to the same objections. As equation 4.4 shows, the energy required for the formation of a critical nucleus involves σ^3, where σ is the surface energy per unit area. This is closely related to the surface tension (per unit length) and is identical with it under simplest conditions. Because of the third power, the nucleation probability is expected to be very sensitive to small changes of σ and this, in turn, is sensitive to contamination, as surface tension always is. Moreover, an analogy with the process of solution is appropriate. For what are probably quite similar reasons, the solution process can be influenced by contaminants in the low parts per million range, as Ives and Plewes (40) have demonstrated. As far as crystal growth in gel is concerned, corresponding experiments have so far been performed only on the calcium tartrate system (41), for which iron is the most effective contaminant. Figure 4.4 shows the results for nucleation probability and contamination. The contamination increases monotonously, as one would expect, and in the course of it the crystals become yellow. (Crystals grown in commercial waterglass are often slightly yellow for the same reason.) Small ferric ion contaminations increase the nucleation probability, whereas larger concentrations inhibit it, with corresponding penalties. The reasons for the reversal in trend are not yet understood.

A control procedure which is free from the above objections involves concentration programming (41). There is, after all, no reason for believing that the reagent concentration which is optimum for nucleation is also optimum for growth. In the course of this procedure, the concentration of the diffusing reagent is initially kept below the level at which nucleation is known to occur. It is then increased in a series of small steps, which can be optimized for any system as regards magnitude and timing. At some stage, as the concentration of the diffusant increases, a few nuclei

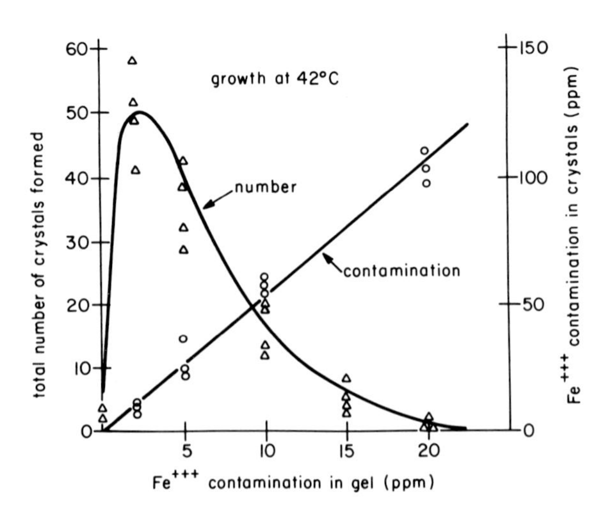

Fig. 4.4.

Effect of iron in the gel on the nucleation and contamination of calcium tartrate crystals [After Dennis and Henisch (41)].

begin to form. It is plausible to assume that these act as sinks and result in the establishment of radial diffusion patterns which actually reduce the reagent concentrations in some of the neighboring locations. In this way, the formation of additional nuclei would be inhibited. Subsequent increases of reagent concentration lead to faster growth but not, in general, to new nucleation. The existing crystals are thus able to grow noncompetitively and their quality is correspondingly good. It has been found empirically that frequent small steps are more beneficial than a few large concentration increases. The method has been successfully applied to the control of nucleation in several systems (see Table below), and has yielded crystals of larger size and of a higher degree of perfection than those produced without programming. The size increase amounts to approximately a factor of 3, except for calcium tartrate, for which nucleation is rare enough, even without programming to reduce the importance of the competitive growth limitation. Figure 1.5b shows a PbI_2 crystal grown by concentration programming, in comparison with specimens (Fig. 1.5a) grown by the ordinary method. It remains to be seen whether the method can be more generally applied, the present indications being that it can. It has, for instance, been used successfully for the nucleation control of antimony sulfur-iodide (SbSI), formed by the reaction of SbI_3 (in HI) and Na_2S (in H_2O). The two reagents are diffused into a U-tube containing an HI gel. Small crystals of BiSI and SbSCl can be grown in analogous ways (42).

Concentration Programming
(Typical Effects of One Form of Procedure on Crystal Size)

Crystal	Typical sizes: largest linear dimensions, mm	
	Without programming	With programming
Calcium tartrate	~ 12	~ 12
Cuprous tartrate	1	3
Lead iodide	3	15
Thallium iodide	0.5	1.5
Calcium carbonate (aragonite spherulites)	0.5	1.5
Cadmium oxalate	2	5

4.5 Nucleation and Gel Structure

Because gels are neither liquid nor solid, there is a great shortage of methods for quantitative structural investigations. As a result, one knows all too little about the relationships between gel structure and nucleation probability. In an attempt to explore these matters, gels have been dialyzed to free them from unwanted reagents, and then rapidly frozen and vacuum dried. By all outward appearances, the gels were not damaged by this procedure, and the silica structure remained essentially intact. Freeze-dried specimens were shadowed with 100–200Å of gold and inspected with a scanning electron microscope. Freezing at liquid nitrogen, dry ice, and ice temperatures gave similar results. Typical structures as reported by Halberstadt and co-workers (39) can be seen in Figures 3.9 and 4.5. The gel is evidently not a simple three-dimensional silica network, as is sometimes supposed. It actually consists of sheet-like structures of varying degrees of surface roughness and porosity, forming interconnected cells. The cell walls are ordinarily curved. The differences between old and new gels, as used above, were too small to show up in these experiments, but the differences between dense and light gels are noticeable (Fig. 4.5a and b).

From pictures of this kind, one can estimate pore sizes, cell dimensions, and cell wall thicknesses. The cell walls seen in dense gels have *pores* from less than 0.1μ to 0.5μ, compared with values from less than 0.1μ to 4μ in low density gels. The cell walls are thicker for the dense gels but it appears that the *cell size* does not depend at all sensitively on gel density. It should be noted that the present pore sizes need not correspond to estimates derived from diffusion experiments, since cell-to-cell diffusion is limited essentially by the pores of smallest diameter encountered along the diffusion path. Figures 3.9 and 4.5 are typical of acid gels (initial pH $\approx$ 5). As one would expect, the pH during gelling has a profound influence on gel structure. As the pH increases the gel structure changes from a distinctly box-like network to a structure consisting of loosely bound platelets which appear to lack cross-linkages; the cellular nature becomes less distinct (Fig. 4.6).

It is also of interest to compare the silica gels with other gels (e.g., gelatin) in which crystals are known to grow. Figure 4.7 shows that the cell walls of gelatin are smooth and relatively free of pores. The cells in gelatin are one order of magnitude larger

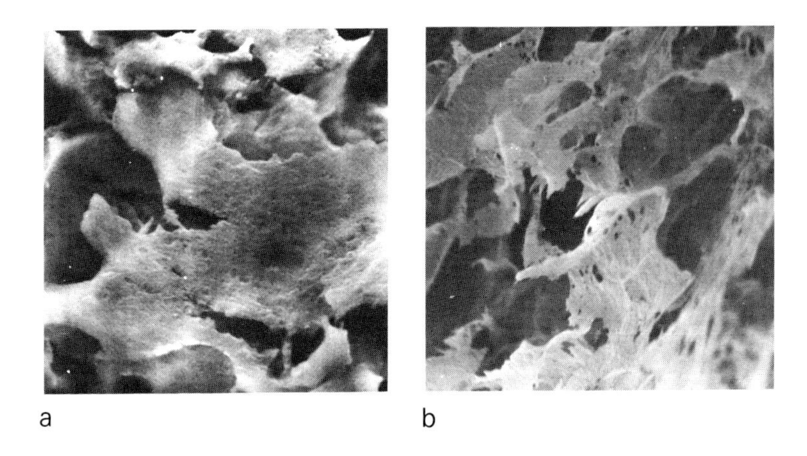

Fig. 4.5.

Structure of dense and light (acid) silica gels (x1500): (a) 0.4 M Na_2SiO_3 in gelling mixture, (b) 0.2 M Na_2SiO_3 in gelling mixture [After Halberstadt, Henisch, Nickl and White (3)].

Fig. 4.6.

Structure of an alkaline silica gel (x2650) [After Halberstadt, Henisch, Nickl and White (39)].

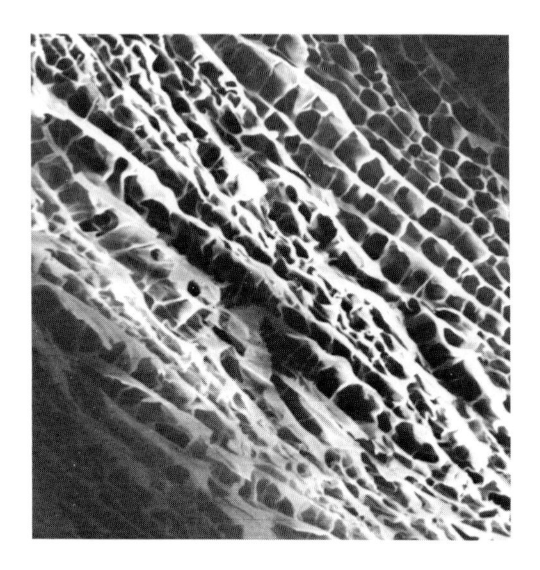

Fig. 4.7.
Structure of gelatin gel (x 265) [After Halberstadt, Henisch, Nickl and White (39)].

than those in silica gel, and this is in harmony with the observation that gelatin supports much more nucleation.

4.6 Crystal Perfection and Distribution

Although only one instance is described in Section 3.1, it is a common observation, made on a great variety of gel growth systems, that the crystals become increasingly scarce and more perfect with increasing distance from the gel interface. Several mechanisms are believed to be at work, singly or jointly, to bring this about.

(a) In cases involving the salts of weak acids, the environment becomes increasingly acid during growth, and the likelihood of a nucleus reaching critical size is correspondingly reduced. It has been shown (29) that calcium tartrate crystals do not ordinarily nucleate at pH values below about 3. Higher perfection at increasing depths would be the natural outcome of reduced competition.

By adding HCl to the $CaCl_2$ solution, fewer and better crystals are produced nearer to the gel interface, which is consistent with the above conclusions. For the reasons given in Section 3.2, similar (but not identical) considerations apply to the growth of PbI_2, even though no strong acid is involved.

(b) It has frequently been noted that the number of crystals growing diminishes with increasing distance from the gel interface. In these regions the diffusion gradients are, of course, different from those near the interface, and this is believed to be the principal explanation of the observed crystal distribution. However, there is a contributory cause arising from gel aging. Crystals which nucleate in the lower regions obviously nucleate in an *older* gel. The existence of such a contribution has since been substantiated by systematic experiments on gels after varying amounts of pre-aging. The results for calcium tartrate and silver iodide are shown in Fig. 4.8. There is the usual statistical spread but the trend, though not spectacular, is significant. As one would expect in the circumstances, it is more pronounced for crystals which nucleate sparingly than for those which nucleate copiously. In lead iodide systems, which come into the latter category, no definite aging tendency is observed. In contrast, aging effects are particularly pronounced for the thin, fragile lead dendrites which may be grown in gels from lead salts on replacement of the lead by zinc. Faust (43) has shown that growth rates are drastically reduced with increasing time after gelling. One possible explanation of aging might be that there is a progressive formation of cross-linkages between siloxane chains, resulting in a gradually diminishing cell size. This, in turn, should lead to a lowering of the nucleation probability, since many potential nuclei, whether homogeneous or heterogeneous, should find themselves in cells of too small a size to support visible growth. By affecting the pore size distribution, an increased number of cross-linkages would also be expected to diminish growth rates.

Gel density should have a similar effect, and in this way the hypothesis can be checked. Figure 4.9 gives a

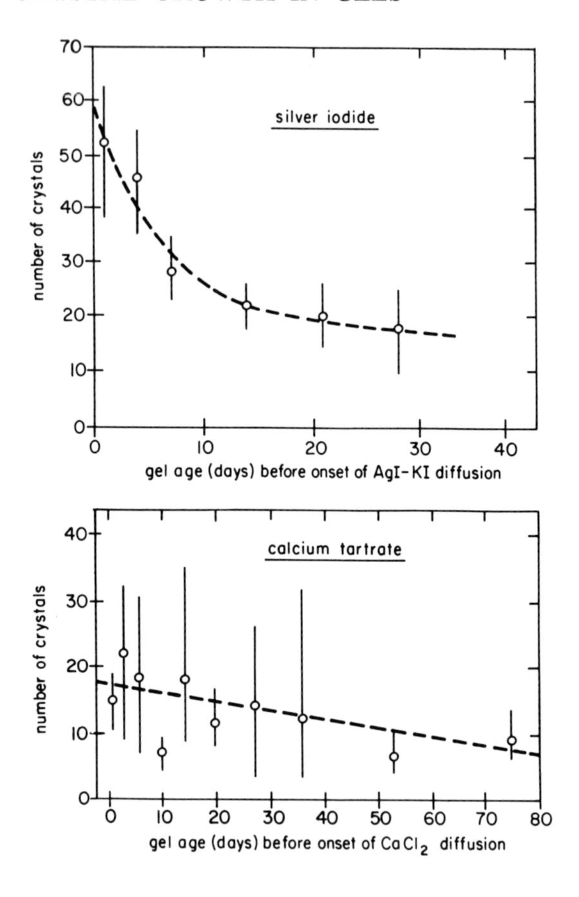

Fig. 4.8.

Effect of gel aging on the nucleation of silver iodide and calcium tartrate; number of crystals growing more than 1 cm from gel surface [After Halberstadt, Henisch, Nickl and White (39)].

comparison of growth results obtained under otherwise identical conditions (pH and reagent concentrations). There is no real evidence that the mechanism whereby the nucleation rate is originally suppressed by the gel is the same as that active during aging, but it is plausible to conclude that it is (39).

(c) The crystals which grow at substantial depths in the gel grow more slowly than those near the top because of the

Fig. 4.9.

Effect of silica gel density on the nucleation of lead iodide, left: 0.1 M Na_2SiO_3 in gelling mixture, right: 0.3 M Na_2SiO_3 in gelling mixture [After Halberstadt, Henisch, Nickl and White (39)].

smaller concentration gradients. There is every reason to believe that the rate of solute arrival affects the perfection of the nuclei formed. Slow diffusion should lead to more perfect nuclei which, because of their higher energy content, should be less likely to reach critical size. This is in agreement with conclusions reached by Sears (44) from considerations relating to the thermodynamics of disordered and possibly anisotropic nucleus formation. The same follows from a minor perturbation of classical nucleation theory by envisaging that imperfect growth boundaries have less interfacial tension. In terms of the greatly oversimplified equation 4.7, this would be seen primarily as a diminution of σ by crystalline disorder. Unlike (a), above, this last mechanism accounts for the crystal distribution without requiring any particular pH change. It also offers an alternative or supplementary interpretation of the "nucleation control by concentration programming" previously described (28).

4.7 Effect of Visible and UV Radiation

Investigations concerning the effect of light on the formation of crystals have an astonishingly long history: Kasatkin (45) quotes references which go back to 1722! Moreover, because conditions during nucleation and growth were not always kept constant with the necessary precision, the results obtained by different workers were often in conflict. Kasatkin reopened the question with reference to the growth (of $NaBrO_3$) from solution, and established a positive effect beyond reasonable doubt. Light increased the growth rate, perhaps through the lowering of a potential barrier at the growth face. Plausible models for such an effect should not be hard to find, but, in the absence of more detailed spectral data, their formulation appears premature.

The first reports of light effects on nucleation (as opposed to growth) in solutions date from 1900, and similar effects in gels have also been reported from time to time (37, 46). For instance, calcium tartrate crystals often grow more prolifically under or after irradiation of the gel systems than in darkness. It is tempting to envisage a new and intriguing quantum process ("photonucleation"), but the available observations lack the necessary generality and do not lend themselves to an interpretation along those lines. Moreover, it was soon found that irradiation of the gel *before* the diffusion process (followed by nucleation in darkness), produced very similar results. The effect was traced to the photo-stimulated development of small CO_2 bubbles, either by dissociation of some tartaric acid or by the release of CO_2 previously dissolved. This may account for the lack of reports to date on light-stimulated nucleation in connection with simpler inorganic gel systems. On the other hand, strong evidence for a genuine light effect on epitaxial growth (and thus heterogeneous nucleation) has recently been reported (47) in the course of vapor growth experiments, and it is not impossible that similar processes may occur in gels. Armington and co-workers (48), for instance, found the illumination appeared to have an adverse effect on the perfection of gel-grown CuCl crystals.

Light may alter the distribution of impurities between the growing crystal and the surrounding gel (37), and that may well be a surface barrier effect as suggested above. It is easily demonstrated with calcium tartrate systems in the presence of iron contamination. Under irradiation by UV or visible light, the iron uptake is inhibited and the resulting crystals are almost colorless; those grown in darkness are yellowish. These matters deserve a great deal of further investigation.

References: Chapter 4

1. Frank, C. F. *J. Faraday Soc.* 5:48 (1949).
2. Frank, F. C. *Phys. Mag.* 42:1014 (1951).
3. Burton, W. K.; Cabrera, N.; and Frank, F. C. *Phil. Trans. Roy. Soc.* 243:299 (1951).
4. Smakula, A. *Einkristalle.* Springer-Verlag, Berlin (1962).
5. Lendvay, E. *Acta Physica Hung.* 17:315 (1964).
6. Tammann, G. *Aggregatzustände.* Barth, Leipzig (1922).
7. Vonnegut, B. *J. Appl. Phys.* 18:593 (1947).
8. Mullin, J. W. *Crystallization.* Butterworths, London (1961).
9. Volmer, M., and Weber, A. *Z. phys. Chem.* 119:225 (1926).
10. Vonnegut, B. *Chem. Rev.* 44:277 (1949).
11. Turnbull, D., and Vonnegut, B. *Industr. Engng. Chem.* 44:1292 (1952).
12. Ostwald, W. *Z. phys. Chem.* 22:289 (1897).
13. VanHook, A. *Crystallization: Theory and Practice.* Reinhold, New York (1963).
14. Jones, W. J. *Z. phys. Chem.* 82:448 (1913).
15. ——— and Partington, J. R. *J. Chem. Soc.* 103:1019 (1915).
16. ——— ———. *Phil. Mag.* 29:35 (1915).
17. Dundon, M. L., and Mack, E. *J. Amer. Chem. Soc.* 45:2479 (1923).
18. Knapp, L. F. *Trans. Faraday Soc.* 17:457 (1922).
19. Koppen, R. *Z. anorg. Chem.* 228:169 (1936).
20. Buckley, H. E. *Crystal Growth.* John Wiley & Sons, New York (1952).
21. Becker, R. *Ann. Phys.* 32:128 (1938).
22. ———. *Proc. Phys. Soc.* 52:70 London (1940).

23. Dunning, W. J. in *Chemistry of the Solid State* (Ed. Garner). Butterworths, London (1955).
24. Amsler, J., and Scherer, P. *Helv. Phys. Acta* 14:318 (1941).
25. ———. *Helv. Phys. Acta.* 15:699 (1942).
26. Kamenetzkaja, D. S. *Rost. Kristallov.* 1:33 (1957).
27. Burgers, W. G. In *The Art and Science of Growing Crystals* (Ed. J. J. Gilman), p. 416. John Wiley & Sons, New York (1963).
28. Henisch, H. K.; Hanoka, J. I.; and Dennis, J. *J. Electrochem. Soc.* 112: 627 (1965).
29. Halberstadt, E. S., and Henisch, H. K. *Proc. Int. Conf. on Crystal Growth,* Birmingham, U. K., July 15–19 (1968); *J. Crystal Growth.* 3, 4:363 (1968).
30. Mott, N. F. *Nature.* 165:295 (1950).
31. Kratochvil, P.; Sprusil, B.; and Heyrovsky, M. *Proc. Int. Conf. on Crystal Growth,* Birmingham, U. K., July 15–19 (1968); *J. Crystal Growth.* 3, 4:360 (1968).
32. Hanoka, J. I. *Polytypism in* PbI_2 *and its Interpretation According to the Epitaxial Theory.* Pennsylvania State University, Thesis (1967).
33. Vand, V., and Hanoka, J. I. *Mat. Res. Bull.* 2:241 (1967).
34. Frank, F. C. *Phil. Mag.* 42:1014 (1951).
35. ———. *Acta Cryst.* 4:497 (1951).
36. Vand, V. *Phil. Mag.* 42:1384 (1951).
37. Henisch, H. K.; Dennis, J.; and Hanoka, J. I. *J. Phys. Chem. Solids.* 26:493 (1965).
38. Perison, J. Pennsylvania State University, personal communication (1968).
39. Halberstadt, E. S.; Henisch, H. K.; Nickl, J.; and White, E. W. *J. Colloid & Interface Science.* 29:469 (1969).
40. Ives, M. B., and Plewes, J. J. *Chem. Phys.* 42:293 (1965).
41. Dennis, J., and Henisch, H. K. *J. Electrochem. Soc.* 114:263 (1967).
42. Dancy, Edna A. Westinghouse Research Laboratories, personal communication (1969).
43. Faust, J. W., Jr. Pennsylvania State University, personal communication (1968).
44. Sears, G. W. In *The Physics and Chemistry of Ceramics* (Ed. Cyrus W. Klingsberg). p. 311. Gordon & Breach, New York (1963).
45. Kasatkin, A. P. *Soviet Physics—Crystallography.* 11:295 (1966).
46. Dennis, J. *Crystal Growth in Gels.* Pennsylvania State University, Thesis, p. 40 et. seq. (1967).
47. Kumagawa, M.; Sunami, H.; Terasaki, T.; and Nishizawa, T. *Jap. J. Appl. Phys.* 7:1332 (1968).
48. Armington, A. F.; DiPietro, M. A.; and O'Connor, J. J. Air Force Cambridge Research Laboratories (Reference 67–0445), Physical Sciences Research Paper No. 334 (July, 1967).

5 Problems Solved and Unsolved

5.1 *Researches on Gel-grown Crystals*

There is ample reason for regarding the study of crystal nucleation and growth as its own justification, but it is reassuring to know that crystals grown in gels have had research applications on a number of occasions. That number is now limited mostly by the availability of suitable specimens, and there is no doubt that intensified activity in gel growth will lead to corresponding rewards in related fields concerned with crystal properties.

Among the researches on crystals grown in gel are the electric spin resonance measurements of Mn^{2+} in calcium tartrate, made by Wakim and co-workers (1). In order to obtain manganese-doped specimens, 0.1% of $MnCl_2$ was added to the $CaCl_2$ solution diffused into the tartaric acid gel (see Section 1.3). The exact Mn concentration in the crystals themselves was not known, but was assumed to be of the same order as in solution. At any rate, changes in the concentration by factors of about 10 produced no change in the resonances detected. As expected, the concentration (as inferred from the measurements) decreased with increased depth of the growing crystals below the liquid interface. The electron resonance measurements were performed at 9100 Mc/s in a microwave spectrometer of conventional type, using magnetic field modulation of 800 c/s. The complete line pattern was found to contain four superimposed spectra. Rotation about the orthorhombic axes showed that these spectra were associated with four distortion axes of the crystalline field, symmetrically dispersed around the b-axis, making angles of $\pm 15°$ with the (001) plane and $\pm 8\frac{1}{2}°$ with the (100) plane, respectively. These axial distortions were detected by observing the relative displacements of the four component spectra for different angular displacements of the crystal. It is possible that the four spectra arise from the four calcium ions in the unit cell. A more complete interpretation was precluded by the fact that the details of the unit cell structure have not yet been worked out. However, the sharpness of the resonance lines was evidence of the high degree of perfection which characterizes crystals grown in this way.

A great deal of work has been done on gel-grown PbI_2, partly with a view to clarifying its complicated band and defect level structure (2–5), and partly in order to add to our understanding of polytypism (6, 7). Both types of investigations had been greatly impeded by the lack of suitable specimens before the advent of the gel method, and major discrepancies had entered the literature. With the gel technique it was found possible to grow not only flat plates, but specimens thick enough to permit measurements of the refractive index in all crystal directions. The measurements showed, among other things, that the dielectric polarization is mainly electronic in character. The material is a photoconductor, with recorded rise and decay times on the order of one or two milliseconds, dependent on the amount of background illumination. A very sharp and dichroic absorption edge can be observed at about 0.525μ which may be explained in terms of a split valence band. There is evidence of impurity levels due to small departures from stoichiometry. Some of the centers present permit radiative recombination, and the measurements have also led to the evaluation of an effective exciton mass. In this general manner, information about the band structure is systematically accumulated.

Research has also been done on important problems of structure. It had been recognized by previous workers that the occurrence of polytypism may depend on the parameters of crystal growth. The gel method offered new possibilities for such studies because the growth conditions could be conveniently varied. In this way, the polytypism in lead iodide crystals has been investigated with a view to finding a link between polytypic sequence and growth conditions (7).

The principal observations on polytypism in gel-grown PbI_2 crystals can be summarized as follows: Slower growth rates favor polytypism (i.e., polytypes $>$ 2H occur in larger proportions), without any pronounced temperature effect. Besides the 2H, 4H, 6R, and 12R polytypes reported by Pinsker, et al. (8), and by Mitchell (6), about 20 new polytypes were found. At all temperatures the percentage of *pure* 2H polytype is lowest for the lowest potassium iodide concentration in the growth systems (see Section 1.3). Independent experiments have shown that the iodine ion concentration determines the growth rate, in agreement with the above conclusion. A similar result has been found by Hagg (9), who showed that slower rates of crystallization from solution produced more polytypism in cadmium iodide. New polytypes conclusively identified by Hanoka and co-workers are 10H, 12H, 14H, 16H, 18H, 24H, 18R, and 24R. Moreover, there may be as many as six types of 12H, two types of 14H, three of 16H, and two of 18H. Polytypes whose identification is still somewhat uncertain are 8H, 32H, 36H, 30R, and 36R. Growth experiments of this general type continue to throw more light upon the problem, but a theory of polytypism which can satisfactorily explain all the observed growth phenomena remains to be found.

In a number of instances, unusual crystals grown in gels have been used for the determination of lattice constants. Bridle and Lomer (10) have done this for cadmium oxalate and copper tartrate, both grown in U-tubes. An alcoholic solution of oxalic acid was used as one of the reagents in the former case. Fair agreement was obtained between measured and calculated densities. Dendrites of metallic lead can be grown with ease and have been used for metallurgical studies by Bedarida (11) and Faust (12), particularly as regards the relationship between growth hillocks, etch pits, and lattice imperfections. The growth of crystals for ferroelectric and laser applications (Section 1.3) is one of the important aims of the gel technique.

5.2 Unsolved Growth and Nucleation Problems

In the instances cited above and presumably in others, the use of the gel method has proved itself as an aid in the solution of research problems. Its further development will no doubt depend on the extent to which its own mechanisms can be clarified and controlled. On this point, many questions remain unanswered, including the all-important one of what one can grow. It appeared at one time that a simple solubility classification could distinguish the crystals that can be successfully prepared in gels, but while early experimenters had faith in this hypothesis, modern workers do not. The notion is full of difficulties, beginning with the meaning of "solubility" in the presence of other compounds and ending with the definition of success. It is necessary only to recall that the growth of one cubic mm of single crystal ZnS in a gel would be regarded as a remarkable success, whereas a calcium tartrate crystal of twenty times this size raises little excitement. Moreover, general experience shows that the crystals which can be grown best are those on which the largest amount of effort has been expended. Casual growth tests are rarely significant, and the great variety of chemical systems involved impedes, for the time being, at any rate, the formulation of generally valid rules.

The most basic questions concern, of course, the all-important nucleation problem. In this respect, the gel method offers a prospect of general rewards, since it appears to permit investigations in the absence of unwanted foreign nuclei. A better understanding of homogeneous nucleation as such will no doubt result, and this, in turn, should reduce the empirical content of gel growth experimentation. In this connection, work on growth systems in which all components are filtered by diffusion through a gel are of greatest importance. At the same time, gel systems offer what appears to be a unique opportunity for experimentation on soluble nucleation centers, possibly of molecular size (Section 4.3).

A second class of problem relates to the chemical role played by internal gel surfaces and the extent to which these are governed by gel structure. The exasperating fact is that different gel media support nucleation and growth to different degrees. A great deal remains to be done to document and explain the detailed nature of these relationships, progress being limited at present by the lack of simpler procedures for the characterization and classification of gels.

A third complex of basic problems is connected with the conditions which govern whether the gel structure will be displaced essentially intact by the growing crystal (as in the case of calcium tartrate), or incorporated into the new solid by crystal growth in the gel interstices (as in the case of calcite). In the former case, a pure gel method can be employed for macro-growth; in the second, its use is limited to the preparation of ordered seeds, to be used in subsequent growth by one of the hybrid processes discussed in Section 3.4. The problem calls for experimentation with sensitive analytic tools applied to crystals grown in carefully controlled gel media.

Apart from these fundamental issues there are a large number of problems which lend themselves well to experimentation at different levels of complexity. These include the diffusion conditions under static and dynamic conditions and their control over ultimate crystal size, the impurity uptake (by itself and in association with gel fragments), crystal perfection and its relationship to growth speed, re-implantation and its dependence on the surface structure of the seed, overdoping and the production of metastable crystals, empirical extension of the experience with different growth systems, the preparation of otherwise intractable crystals, and crystal growth in nonaqueous gels, to mention only a few. In each case, at least some significant advances could be made by very simple means, just as they were in the early days of formal experimentation with growth in solution. For once, the scales are not completely weighed in favor of the deluxe experimenter, and there are not many fields left in the physical sciences which provide self-funded amateurs and unfunded professionals alike with such a beautiful, congenial, and useful field of endeavor.

References: Chapter 5

1. Wakim, F. G.; Henisch, H. K.; and Atwater, H. A. *J. Chem. Phys.* 42:2619 (1965).
2. Henisch, H. K., and Srinivasagopalan, C. *Solid State Communications.* 4:415 (1966).
3. Dugan, A. E., and Henisch, H. K. *J. Phys. Chem. Solids.* 28:1885 (1967).
4. ________. *J. Phys. Chem. Solids.* 28:971 (1967).
5. ________. *Phys. Rev.* 171:1047 (1968).
6. Mitchell, R. S. *Z. Krist.* 111:372 (1959).
7. Hanoka, J. I.; Vedam, K.; and Henisch, H. K. In *Crystal Growth* (Ed. S. Peiser). Pergamon Press, New York (1967).
8. Pinsker, Z. G.; Tatarinova, L.; and Novikova, V. *Acta Physio-chim. URSS.* 18:378 (1943).
9. Hagg, G. *Colloq. Intern. Centre Natl. Rech. Sci., Réactions dans L'état Solide.* 10:5 (1948).
10. Bridle, C., and Lomer, T. R. *Acta Cryst.* 19:483 (1965).
11. Bedarida, F. *Periodico di Mineralogie.* 33:1 (1964).
12. Faust, W., Jr. Pennsylvania State University, personal communication (1969).

Supplementary Notes

Mercuric Iodide

One of the earliest records: Holmes: *J. Phys. Chem.* 21:709 (1917). Mercuric chloride diffused into silica gel containing potassium iodide. Needle-shaped crystals formed, sometimes yellow at first, then red. More recent work by Averett: *Sci. Amer.* 206:155 (1962); and Kurz: *Ohio J. Science.* 66:198 (1966).

Arsenates

Growth described by Keester (personal communication). Acetic acid silica gels contained 0.5 M $NaHAsO_4$. Various supernatant cation solutions. Lead arsenate (schultenite) obtained by diffusing 1 M lead acetate, copper arsenate (olivenite) by diffusing 1 M copper nitrate. Growth fast (2 weeks) in former case, slow (several months) in latter.

Gold

One of the earliest records by Hatschek and Simon: *J. Soc. Chem. Ind.* 31:439 (1912); and *Kolloid Zeitschrift* 10:265 (1912). Reduction of gold salts by the use of oxalic acid, ammonium formate, ferrous sulfate, sodium sulfate, carbon monoxide, sulfur dioxide, hydrogen, and ethylene. Preparation also described by Holmes: *Colloid Chemistry,* p. 796, Chemical Catalog Co., New York (1926). Example: equal volumes of 1.06 density waterglass and 1.5 M sulfuric acid mixed; 1 cc of 1% gold chloride solution added to 25 cc of acid gelling mixture; gel covered with 8% oxalic acid solution. Liesegang rings of microcrystalline gold produced, with a few larger crystallites scattered between them.

Lead Chloride

Growth reported by Hatschek: *Kolloid Zeitschrift* 8:193 (1911). Needles of 15 mm length obtained in silicic acid gels, shorter in gelatin or agar.

Lead

One of the earliest records by Simon: *Kolloid Zeitschrift* 12:171 (1913). Gel contained lead acetate and a lump of metallic zinc. Method discussed by Holmes [in *Colloid Chemistry.* p. 796. Chemical Catalog Co., New York (1926)], who quotes 0.01 M as an appropriate acetate concentration.

Copper

One of the earliest records by Holmes: *Colloid Chemistry,* p. 796, Chemical Catalog Co., New York (1926). Perfect tetrahedra obtained, using a 1% hydroxylamine hydrochloride solution to reduce copper sulfate incorporated in the gel.

Nickel and Cobalt Phosphates

Growth experiments described by Kurz: *Ohio J. Science* 66:349 (1966). Results not similar for the two materials, despite the frequently isomorphous nature of Ni and Co salts.

Silver Sulfate

One of the earliest records by Holmes: *Colloid Chemistry,* p. 796, Chemical Catalog Co., New York (1926). Very fast growth in waterglass containing 1.5 M sulfuric acid, with 1 M silver nitrate solution diffusing.

Zeolites

Crystals grown in acrylic gels by Ciric: *Science* 55:689 (1967) to sizes of the order of 100μ. Gels prepared by dissolving 5 g of "Carbopol 934 (2)," an acrylic acid polymer, in 100 cc of 1 M NaOH. Aqueous slurries of sodium aluminate on one side and sodium metasilicate on the other served as reagents. A "Tygon" tube was used as container, with a diffusion path of 30 cm.

Mercuric Chloride

Basic form $HgCl_2 \cdot 2HgO$, grown by Kurz: *Ohio J. Science* 66:284 (1966), who describes an investigation of the factors which control growth results.

Index